ENGINEERING DRAWING AND DESIGN II

Peter Astley

T. Eng. (CEI), M.I. Plant E., M.I.E.D.

Foley College of Further Education, Stourbridge

M

First published 1978 by
THE MACMILLAN PRESS LTD
London and Basingstoke
Associated companies in Delhi Dublin
Hong Kong Johannesburg Lagos Melbourne
New York Singapore and Tokyo

Printed in Great Britain by
A. Wheaton & Co. Ltd.,
Exeter, Devon

British Library Cataloguing in Publication Data

Astley, Peter George Warren
    Engineering drawing and design II. —(Macmillan
technician series).
    1. Engineering drawing 2. Mechanical drawing
    I. Title II. Series
    604′.2′4   T353

ISBN 0–333–21703–9

ENGINEERING DRAWING AND DESIGN II

Good design is usually achieved when persons with suitable education and training can collect and simplify information relevant to a requirement; make the information manageable within the available resources of materials and techniques, and ultimately user satisfaction will be obtained.

Contents

Foreword

This book is written for one of the many technician courses now being run at technical colleges in accordance with the requirements of the **Technician Education Council** (TEC). This Council was established in March 1973 as a result of the recommendation of the Government's Haslegrave Committee on Technical Courses and Examinations, which reported in 1969. TEC's functions were to rationalise existing technician courses, including the City and Guilds of London Institute (C.G.L.I.) Technician courses and the Ordinary and Higher National Certificate courses (O.N.C. and H.N.C.), and provide a system of technical education which satisfied the requirements of 'industry' and 'students' but which could be operated economically and efficiently.

Four qualifications are awarded by TEC, namely the Certificate, Higher Certificate, Diploma and Higher Diploma. The **Certificate** award is comparable with the O.N.C. or with the third year of the C.G.L.I. Technician course, whereas the **Higher Certificate** is comparable with the H.N.C. or the C.G.L.I. Part III Certificate. The **Diploma** is comparable with the O.N.D. in Engineering or Technology, the **Higher Diploma** with the H.N.D. Students study on a part-time or block-release basis for the Certificate and Higher Certificate, whereas the Diploma courses are intended for full-time study. Evening study is possible but not recommended by TEC. The Certificate course consists of fifteen Units and is intended to be studied over a period of three years by students, mainly straight from school, who have three or more C.S.E. Grade III passes or equivalent in appropriate subjects such as mathematics, English and science. The Higher Certificate course consists of a further ten Units, for two years of part-time study, the total time allocation being 900 hours of study for the Certificate and 600 hours for the Higher Certificate. The Diploma requires about 2000 hours of study over two years, the Higher Diploma a further 1500 hours of study for a further two years.

Each student is entered on to a **Programme** of study on entry to the course; this programme leads to the award of a Technician Certificate, the title of which reflects the **area of engineering** or science chosen by the student, such as the Telecommunications Certificate or the Mechanical Engineering Certificate. TEC have created three main **Sectors** of responsibility

Sector A responsible for General, Electrical and Mechanical Engineering

Sector B responsible for Building, Mining and Construction Engineering

Sector C responsible for the Sciences, Agriculture, Catering, Graphics and Textiles.

Each Sector is divided into Programme committees, which are responsible for the specialist subjects or programmes, such as A1 for General Engineering, A2 for Electronics and Tele-communications Engineering, A3 for Electrical Engineering, etc. Colleges have considerable control over the content of their intended programmes, since they can choose the Units for their programmes to suit the requirements of local industry, college resources or student needs. These Units can be written entirely by the college, thereafter called a college-devised Unit, or can be supplied as a Standard Unit by one of the Programme committees of TEC. **Assessment** of every Unit is carried out by the college and a pass in one Unit depends on the attainment gained by the student in his coursework, laboratory work and an end-of-Unit test. TEC moderate college assessment plans and their validation; external assessment by TEC will be introduced at a later stage.

The three-year Certificate course consists of fifteen Units at three Levels: I, II and III, with five Units normally studied per year. A typical programme might be as follows.

Year I	Mathematics I	Standard Unit	
	Science I	Standard Unit	
	Workshop Processes I	Standard Unit	six Level I Units
	Drawing I	Standard Unit	
	General and Communications I	College Unit	
Year II	Engineering Systems I	College Unit	
	Mathematics II	Standard Unit	
	Science II	Standard Unit	
	Technology II	Standard Unit	

	General and Communications II	College Unit	six Level II Units
Year III	Industrial Studies II	College Unit	
	Engineering Systems II	College Unit	
	Mathematics III	Standard Unit	
	Science III	Standard Unit	three Level III Units
	Technology III	College Unit	

Entry to each Level I or Level II Unit will carry a prerequisite qualification such as C.S.E. Grade III for Level I or O-level for Level II; certain Craft qualifications will allow students to enter Level II direct, one or two Level I Units being studied as 'trailing' Units in the first year. The study of five Units in one college year results in the allocation of about two hours per week per Unit, and since more subjects are often to be studied than for the comparable City and Guilds course, the treatment of many subjects is more general, with greater emphasis on an **understanding** of subject topics rather than their application. Every syllabus to every Unit is far more detailed than the comparable O.N.C. or C.G.L.I. syllabus, presentation in **Learning Objective** form being requested by TEC. For this reason a syllabus, such as that followed by this book, might at first sight seem very long, but analysis of the syllabus will show that 'in-depth' treatment is not necessary—objectives such as '. . . states Ohm's law . . .' or '. . .lists the different types of telephone receiver . . .' clearly do **not** require an understanding of the derivation of the Ohm's law equation or the operation of several telephone receivers.

This book satisfies the learning objectives for one of the many TEC Standard Units, as adopted by many technical colleges for inclusion into their Technician programmes. The treatment of each topic is carried to the depth suggested by TEC and in a similar way the **length** of the Unit (sixty hours of study for a full Unit), **prerequisite qualifications, credits for alternative qualifications** and **aims of the Unit** have been taken into account by the author.

Preface

This book has been written to satisfy the aims and Unit topic areas of the Technician Education Council Standard Unit, Engineering Drawing and Design II, Unit Level II. It is assumed that the student will already possess a pass grade at G.C.E. O-level or a C.S.E. Grade I in engineering or technical drawing; therefore a thorough coverage of basic draughting techniques has not been attempted.

Having trodden the path from junior draughtsman to chief draughtsman, I sincerely hope that this publication will help to encourage students to further their studies. Design can mean different things to different people, but the content of this TEC Unit affords the opportunity for lecturers to provide a real-world approach to the subject.

For checking the script, my thanks go to Mr R. F. Hall, M.I.E.D. I am also grateful to various companies for permission to use photographs; their names appear within the appropriate sections of the book.

Finally, I should like to thank my wife, Sheila, for her steadfast work in preparing the script for publication and for showing her usual patience.

Finchfield, 1977   PETER ASTLEY

1 Engineering Communication

THE ENGINEERING DRAWING

Consider the quantity of drawings required for the manufacture of an aircraft or a car. The graphical language of draughting is used throughout the world as a means of expression for the engineering designer and draughtsman in helping to create such products. A basic idea can commence with a rough sketch, and is then refined until it is eventually in the form of a finished orthographic drawing. Drawings are required for products, and drawings are required for the manufacture of the machines that make the products. Production aids, such as jigs, fixtures and tools, all begin life on the drawing board.

How Much Draughting is Needed?

As a student preparing for the engineering profession you may question how much draughting you should accomplish during the learning stage. If you wish to be a scientist you will need less draughting skill than if you intend to become an engineer. On the other hand if your aim is to be a designer, then the more draughting knowledge and skill you possess the greater will be your initial success in this area. All technicians should have a thorough understanding of the following basic areas of draughting.

(1) Geometric construction
(2) Orthographic projection
(3) Dimensioning
(4) Freehand sketching
(5) Sectional views
(6) Auxiliary views
(7) Fastening devices
(8) Working drawings
(9) Pictorial drawings

To gain admittance to the course for which this book is written, you will have obtained a certain knowledge of some of these areas through study of C.S.E., G.C.E. or other engineering drawing courses and examinations.

As an engineering draughtsman you will be required to produce sketches, detail drawings, general arrangement drawings and parts lists. You will be the all-important link in the communication chain between the designer and the manufacturing section. Manufacturing personnel are paid to produce components to your requirements. They are not paid to spend their time sorting out and querying drawing errors or lack of manufacturing information. A correctly sketched detail on the back of a canteen menu card is of more value than an incorrect detail prepared in all other respects to BS 308.

Suppose that you have been requested to make manufacturing drawings for replacement parts on a machine used in the toolroom in your organisation. If you are suitably prepared with pad, pencil, rule, etc., it may be convenient or necessary for you to obtain all the information you require on the shop floor without taking the parts back to the drawing office. Neat freehand sketches will enable you to prepare drawings in the office, without the need for further visits to the workshop, which in any event cause disruption to production (see figure 1.1). You can now proceed to prepare detail drawings of the various parts, provide an arrangement drawing showing the parts in position, and finally compile the parts list (see figures 1.2 and 1.3).

However, let us assume that discussion is required, with the personnel involved, regarding the proposed replacement of parts. Neat sketches giving full information with regard to requirements will often suffice at this stage, and it is as well to remember that some of the people taking part in the discussion may not necessarily be technical personnel. Your communication skills can therefore be stretched even at this preliminary stage of drawing. How many parts are there to this assembly? How do they fit? Is it complicated? How much work does it entail? These and many more questions can be asked at technical meetings. A photostat copy given to each member of the meeting, with the layout shown, will help intercommunication considerably.

Simplicity of presentation is the keynote. It is not your task to make a simple nut and bolt a complicated affair. Your tidy presentation of draughtsmanship is an important step that can help you to make positive engineering decisions at a later stage in your career. Even old sayings such as 'bad printing spoils a good drawing, but good printing improves any drawing' are worth considering at this stage. A few well-chosen words with a neat

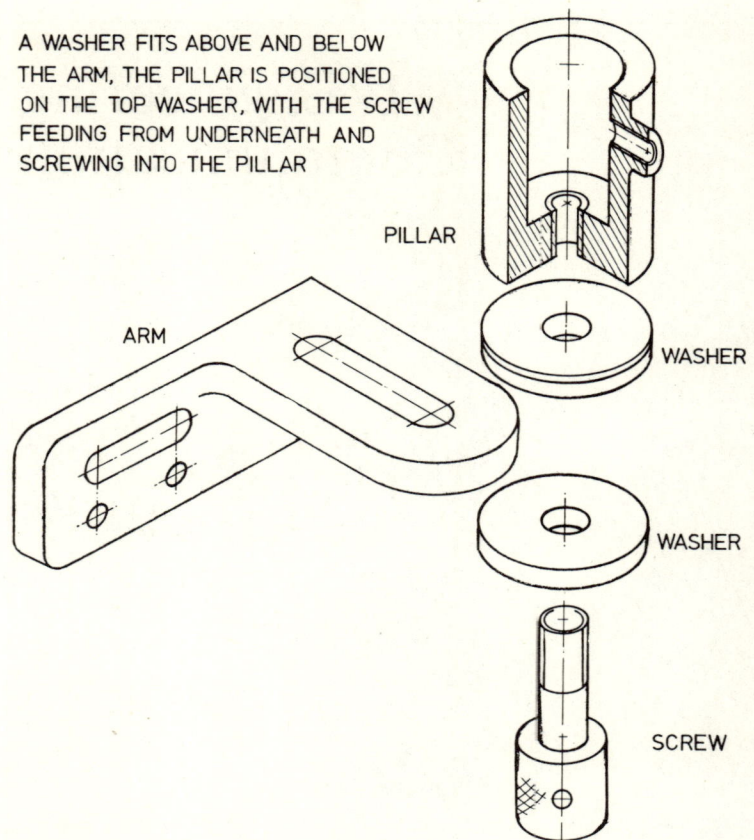

A WASHER FITS ABOVE AND BELOW THE ARM, THE PILLAR IS POSITIONED ON THE TOP WASHER, WITH THE SCREW FEEDING FROM UNDERNEATH AND SCREWING INTO THE PILLAR

Figure 1.1

sketch will tell even the non-technically minded person how the parts fit together.

For example, in figure 1.1 one could say in note form on the sketch, 'a washer fits above and below the arm, the pillar is positioned on the top washer, with the screw feeding from underneath and screwing into the pillar'. Shapes, dimensions and assembly sequence are clear to all concerned. It is a simple example of engineering communication, and does away with other less clear-cut methods of presentation which could be, and often are, provided. You will also need to specify quantities, materials, heat treatment and surface finish as part of your task. Your decisions will affect buying, stock control, accounts, inspection and other departmental procedures.

Engineering drawings should comply with British Standard 308: Engineering Drawing Practice. This standard is in three parts and the Students' Edition contains the whole of BS 308: Part 1: 1972

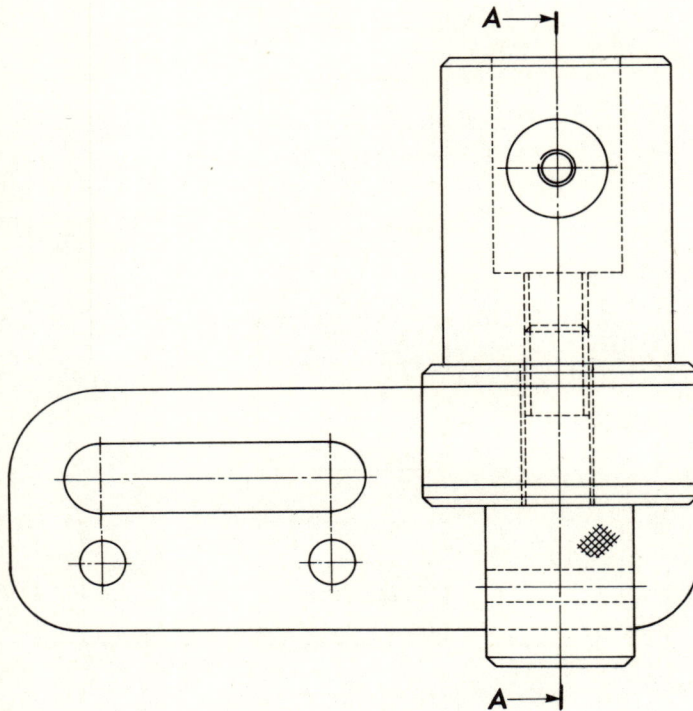

and selected pages from Parts 2 and 3. Every drawing office should have a copy of the complete standard, and students who intend to become draughtsmen should consider purchasing the Students' Edition. Part 1 of the standard covers drawing layout, types of drawing, lines and linework, lettering, projection, views on drawings, sections, conventional representation, scales and abbreviations for use on drawings.

Drawings must be completed in either *first angle* (European) or *third angle* (American) orthographic projection. Both systems are

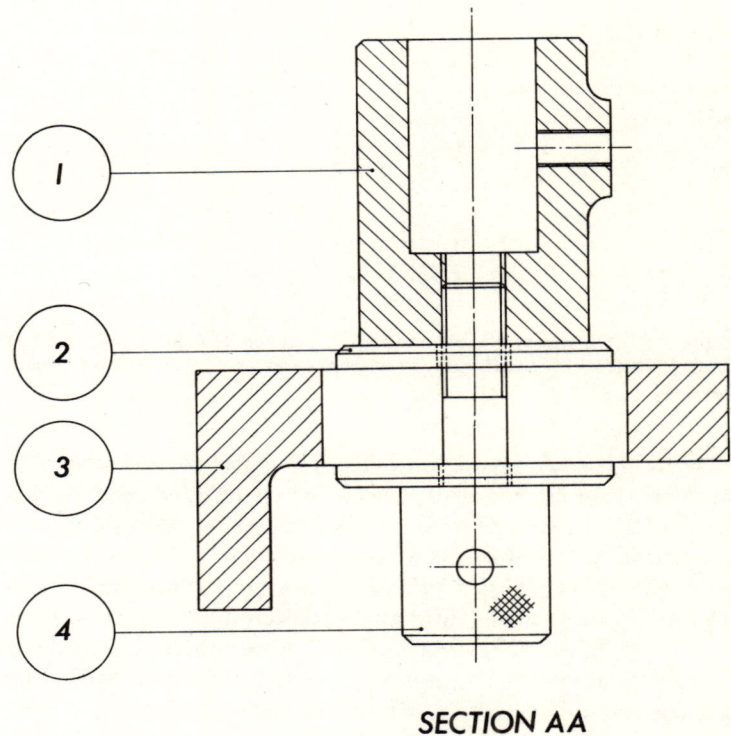

Figure 1.2 Consider that the elevation and sectional elevation are approximately full size. Choose a suitable 'A' size sheet, prepare the layout as shown on the standard sheet and copy the given views. Detail the parts on separate small sheets; complete a separate parts list. Remember that a bold outline, neatly laid-out views and a clear style of printing are basic requirements for any drawing

Figure 1.3

SECTION AA

approved internationally and are regarded as being of equal status. Individual companies use the system of their choice. The projection system used on a drawing should be indicated by the appropriate symbol and suitably positioned (see figures 1.4 and 1.5).

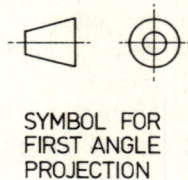

SYMBOL FOR
FIRST ANGLE
PROJECTION

Figure 1.4

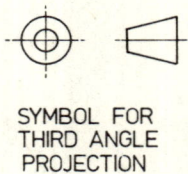

SYMBOL FOR
THIRD ANGLE
PROJECTION

Figure 1.5 The system of projection used on a drawing should be indicated by a symbol. Alternatively the direction in which the views are taken should be clearly indicated

With first angle projection each view shows what is seen when looking on the far side of an adjacent view. Consider a simple angle bracket (see figure 1.6) in first angle projection. With third angle projection each view shows what is seen when looking on the near side of an adjacent view. Consider again a simple angle bracket (see figure 1.7), this time in third angle projection.

The number and choice of views should be chosen to provide the maximum amount of information clearly. Most companies have pre-cut and pre-printed drawing sheets, with title block and other relevant columns. Each detail may be drawn on a separate sheet, or the arrangement and the details may be on a single sheet. Certain accepted methods, appertaining to component features, typical of BS 308 will be outlined in the following pages. With this in mind we can now proceed to complete detail drawings, and an arrangement

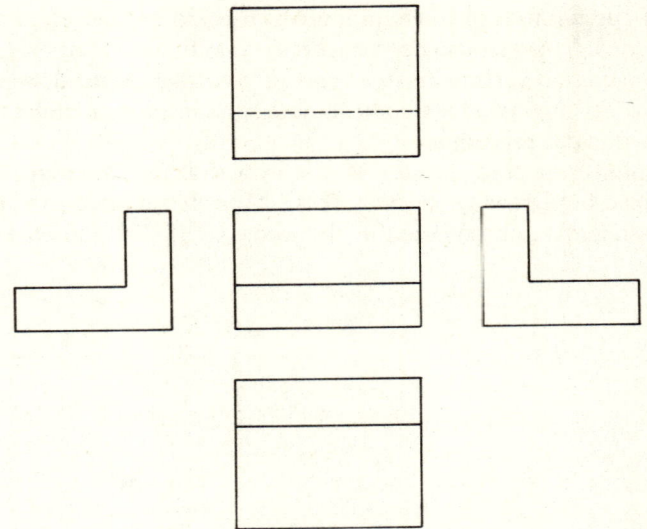

Figure 1.6

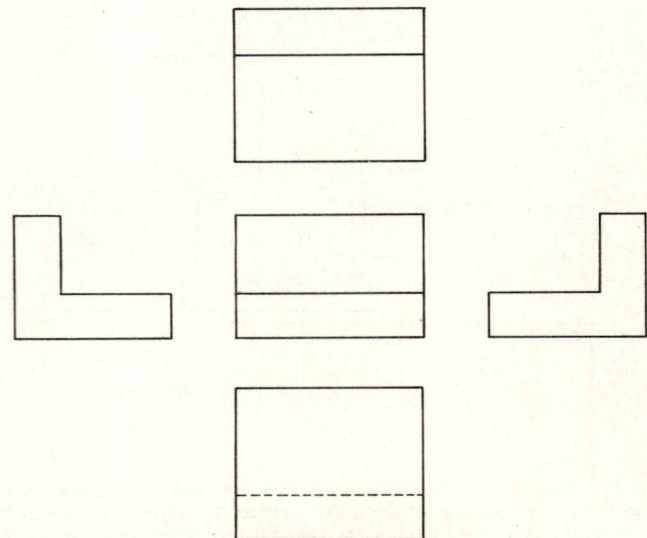

Figure 1.7

drawing, of the special tool support parts in first angle projection. We shall also prepare a parts list.

One consideration before proceeding with the task in hand: thought must be given to the accumulation of error due to incorrect methods of presenting dimensions. The following simple example should be understood. The need for the correct approach, now shown, will appear time and time again throughout your drawing office career!

It is essential that engineering drawings are dimensioned so that the information provided by the draughtsman can be interpreted in one way only. No doubt must be left, in the mind of the fitter, toolmaker or machinist, regarding the dimensions given. It can be as misleading to provide too many dimensions as too few dimensions on a drawing. Furthermore, bad dimensioning can lead to the accumulation of errors just mentioned. Therefore, before considering a specific problem, a note commonly found on an engineering drawing must be understood. The note can read

ALL DIMENSIONS TO BE $\pm$ 0·4 mm

UNLESS OTHERWISE STATED.

This simply means that a stated nominal dimension can be exceeded by 0.4 mm, or that it can be made less by 0.4 mm. These are the 'high' and 'low' limits of size. Any figure between these limits is acceptable.

Example—A headed pin is dimensioned 32 ± 0.3 underhead. State the maximum and minimum sizes allowed for manufacture (see figure 1.8)

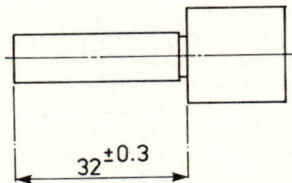

Figure 1.8

the *nominal* size is 32 mm, 0.3 mm is allowed above nominal, so *maximum* size = 32 + 0.3 = 32.3 mm

the *nominal* size is 32 mm, 0.3 mm is allowed below nominal, so *minimum* size = 32 − 0.3 = 31.7 mm

All limits must be chosen with care. Fine limits increase the cost of manufacture and coarse limits can affect interchangeability of mass-produced parts. However, our concern here is the correct dimensioning of engineering drawings, the subject of limits and fits being a separate consideration.

Let us now consider the case of accumulated error arising from the dimensioning of a machined block. The distance between the *top face* B and *bottom face* A depends upon five dimensions, that is, three on the left and two on the right. Let us show, by simple calculations, the difficulties arising from accumulated error due to this *excess* of dimensions (see figure 1.9).

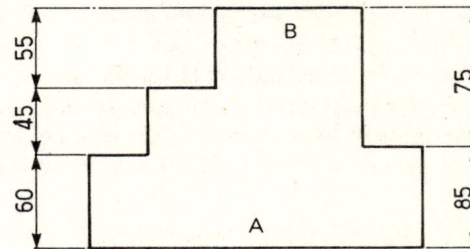

ALL DIMENSIONS
TO BE $\pm$ 0.1mm

Figure 1.9

Step 1   Calculate the distance from A to B using maximum limits of size for the dimensions on the left-hand side
    60.1 + 45.1 + 55.1 = 160.3 mm
Step 2   Calculate the distance from A to B using maximum limits of size for the dimensions on the right-hand side
    85.1 + 75.1 = 160.2 mm

Compare this answer with the result obtained in the Step 1. Which is correct?

Step 3   Calculate the distance from A to B using minimum limits of size for the dimensions on the left-hand side

$$59.9 + 44.9 + 54.9 = 159.7 \text{ mm}$$

Compare this answer with the result obtained in the Step 2.

Step 4   Calculate the distance from A to B using the minimum limits of size for the dimensions on the right-hand side

$$84.9 + 74.9 = 159.8 \text{ mm}$$

Compare this answer with the results obtained in Steps 1 and 3. Again the sizes do not agree.

Even if we agree to accept the dimensions calculated from the right-hand side, we have allowed the distance AB to be 0.2 mm greater or less than its nominal value. This conflicts with the general instruction and is not acceptable.

We have now established that bad dimensioning leads to accumulated errors and that too many dimensions lead to inconsistency. Consider the same component, but dimensioned as follows (see figure 1.10). By using face A as a datum we have avoided all these difficulties and yet have still kept to the limits of ± 0.1 mm that have been specified. No greater demands have been made on the manufacturing process, and the distance from A to B is governed by its own limits of size only, that is, its finished size must be between 160.1 mm and 159.9 mm. By dimensioning from a datum, each dimension is considered separately and does not depend upon other dimensions for the positioning of features on a component.

With the knowledge that all manufactured features must be produced to within specified limits, it is worth while considering the implications of 'chain dimensioning' when applied to gearbox, radio and television, or aircraft and ship design, with all the many separate details of parts involved.

The separate parts list shown (see figure 1.11) is typical of that used in a manufacturing organisation. It is prepared in the drawing office and is issued to the various departments that will be involved with the product. The buying department will order any items or materials not in stock from outside suppliers; the stores

ALL DIMENSIONS
TO BE ± 0.1mm

Figure 1.10

CUSTOMER			PARTS LIST		COMPILED DATE		ORDER No.	
CUSTOMER'S ORDER No.								
DESCRIPTION					CHECKED DATE		G.A.DRG No.	
DELIVERY DUE								
ITEM No.	QUANTITY	DESCRIPTION	MATERIAL	DRG No.	SUPPLIER	ORDER No.	DELY REQd	
X Y Z MANUFACTURING COMPANY LIMITED					SHEET No. of SHEETS			

Figure 1.11 The separate parts list, issued from the drawing office to all departments concerned. A customer who places an order for work to be completed by an outside manufacturing unit will have an order number for use within the company—a Customer's Order Number. On receipt of the order, the manufacturing company will provide its own order number for use within its system (top right-hand side of sheet). The manufacturing company may, in turn, subcontract to a further outside supplier, and will issue a further order number for parts manufactured (see column 7). With a large assembly, several parts lists may be required, so there will be a column to read, for example, 'Sheet No. 5 of 9 Sheets'. The list provides a further example for the need for correct communication in an industrial organisation

department will require the information for stock records; and the estimating department will need the information for costing. An error will therefore provide little popularity for the person responsible for compiling the list! As an exercise, try to think of other departments who may need the parts list.

Figure 1.12 Typical layout of a detail drawing sheet (used for small details). For further examples see BS 308: 1972

A small detail format is shown on a blank sheet (see figure 1.12). Complete the column headings as on a typical sheet and then draw the detail of the holding screw. Complete all the other details in a similar manner. Prints of the original drawing should then be taken and issued to all concerned. The student will now appreciate how paperwork accumulates in every department. This affects the number of staff employed, floor space and management levels, and in no time at all the back-street firm is a multi-national corporation!

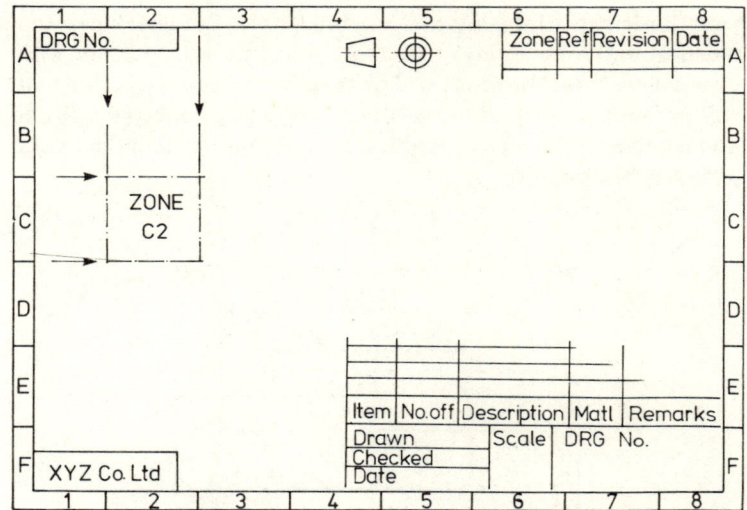

Figure 1.13 Typical layout of a standard drawing sheet (not to scale). For further examples see BS 308: 1972. Note (1) the grid reference, which locates the zone and enables revisions to be located quickly and accurately; (2) the revision table, which provides a complete record of modifications made to the drawing; (3) the drawing number, which appears in opposite corners for quick reference for filing; (4) the method of projection; (5) the provision of a parts list in the case of an assembly drawing—as with the revision table, the parts list continues inwards to allow additions

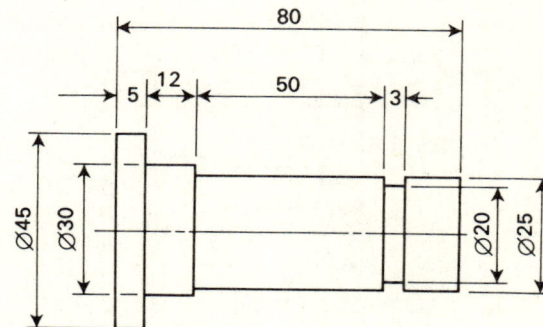

Figure 1.14 Staggered dimensions to avoid confusion

Typical details showing neatness of layout are provided (see figures 1.14 to 1.37). The method in which you present your work is your trademark. Develop a clear, bold style. Do not worry too much about your speed of presentation; this will increase with experience. Your drawings will be used by internal workshops or will perhaps be sent to outside manufacturing concerns both at home and abroad. Your graphical representation can affect your company's reputation.

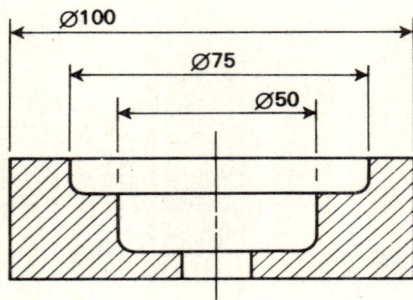

Figure 1.15 Overall dimension placed outside the intermediate dimensions, with one intermediate dimension omitted

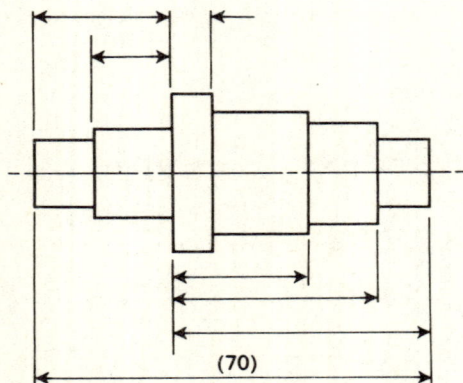

Figure 1.17 A further example of auxiliary dimensions. It should be noted that auxiliary dimensions do not govern machining operations in any way

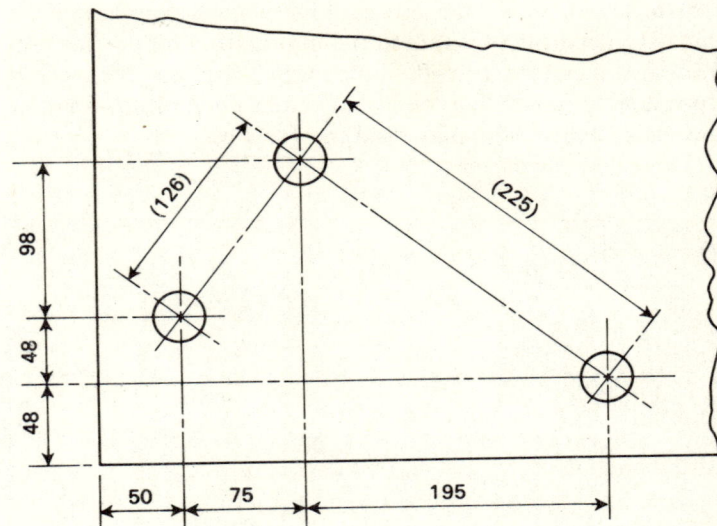

Figure 1.16 Overall dimensions given as an auxiliary dimension and therefore shown in parentheses. This applies where all the intermediate dimensions shown are really necessary

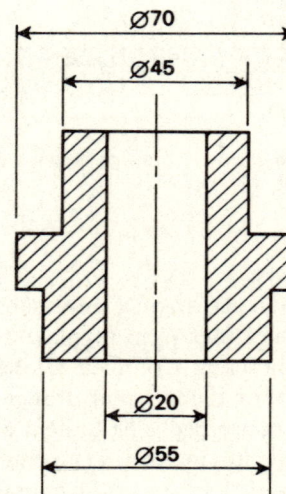

Figure 1.18 Indicating diameters; the abbreviation 'DIA' should be used only in note form

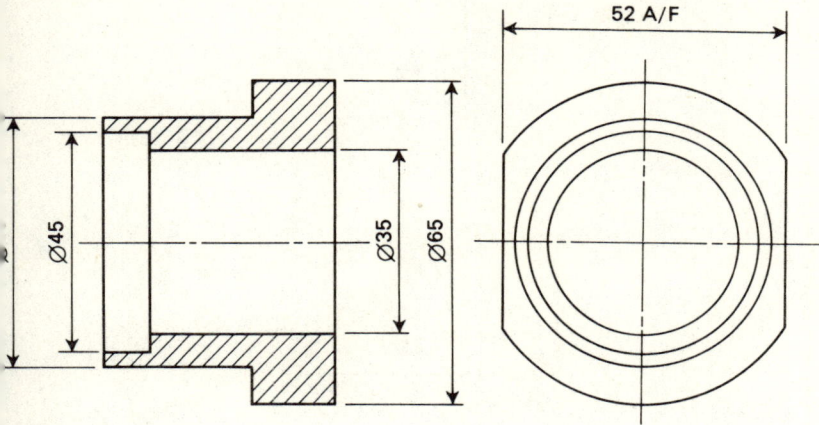

Figure 1.19   Diameter dimensions placed on the most convenient view to ensure clarity. The example shows such dimensions on a longitudinal view rather than the 'concentric circles' end view

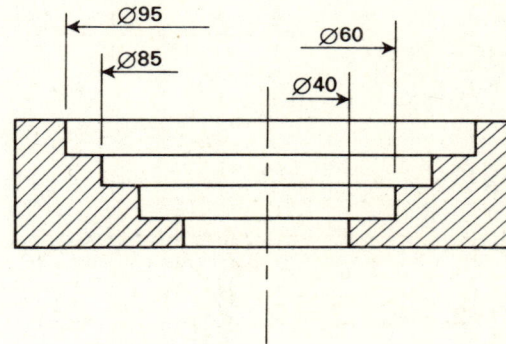

Figure 1.21   Method of dimensioning when space is restricted

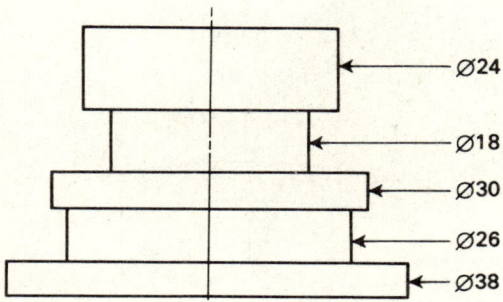

Figure 1.20   Leaders may be used

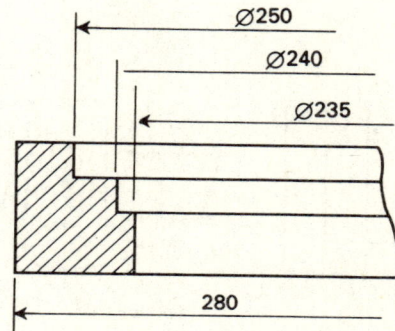

Figure 1.22   Method of dimensioning when space is restricted

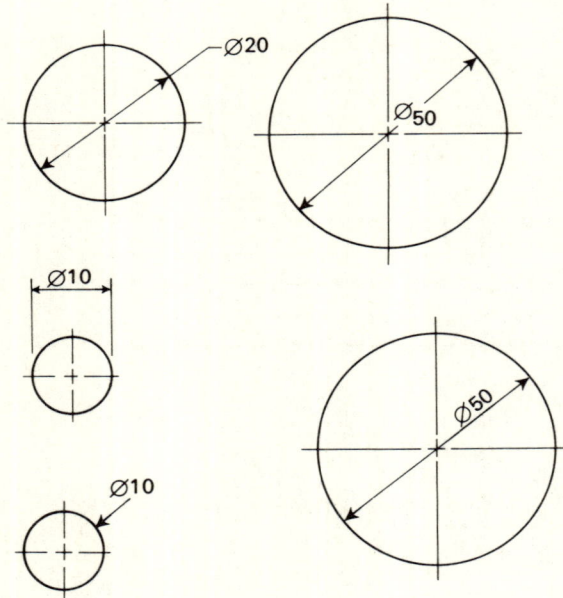

Figure 1.23   The dimensioning of circular features

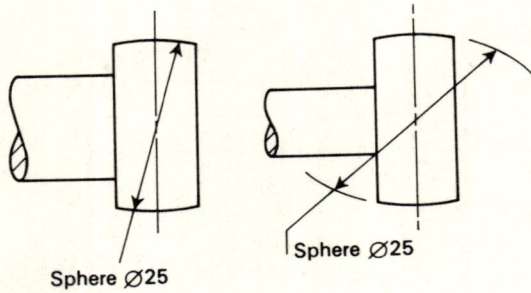

Figure 1.24   Spherical diameters

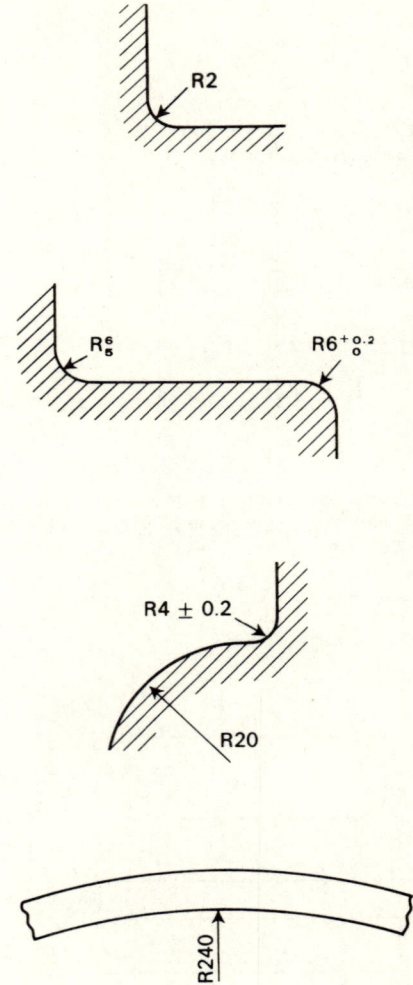

Figure 1.25   Radii—note that the symbol R is used in capital form preceding the dimension

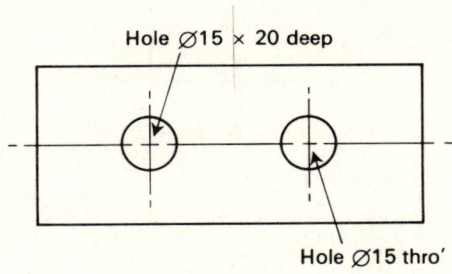

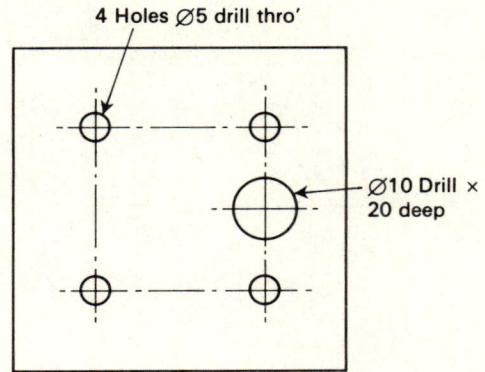

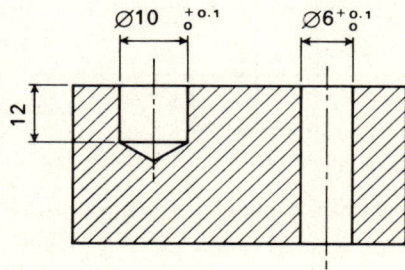

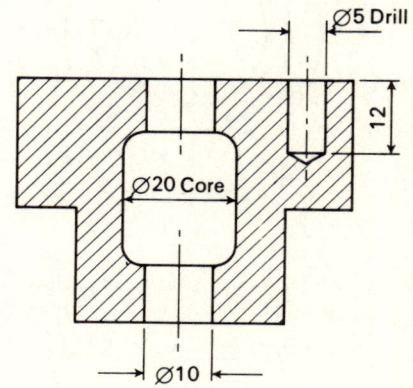

Figure 1.26   Size of holes

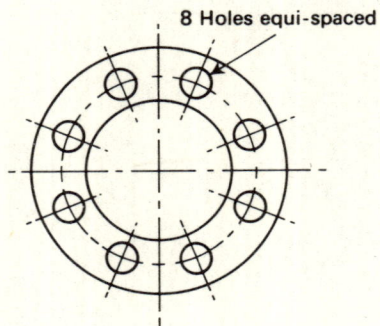

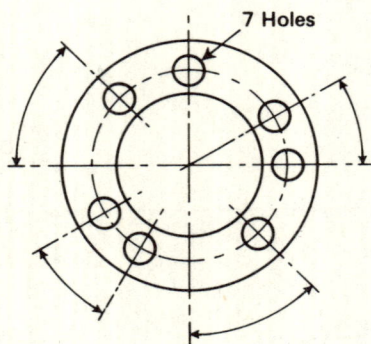

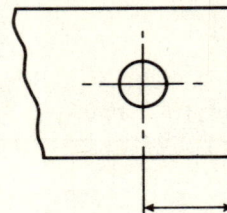

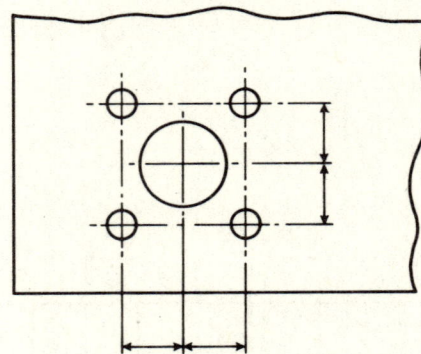

Figure 1.27   Positioning of holes and features on pitch circles

Figure 1.28   Positioning of holes using rectangular coordinates

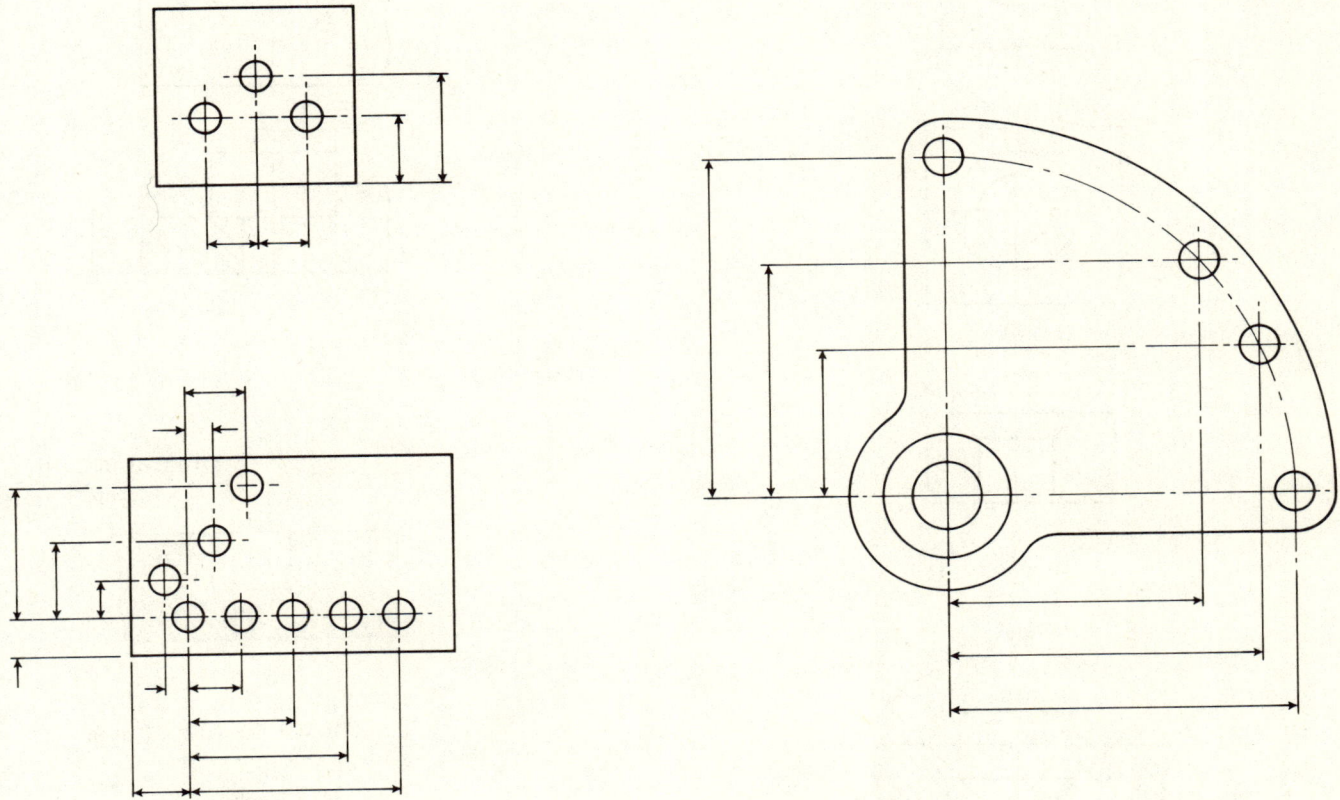

Figure 1.28 *(continued)*

Ø5 C'SK at 90° to Ø12

90°
Ø15

90°
Ø15

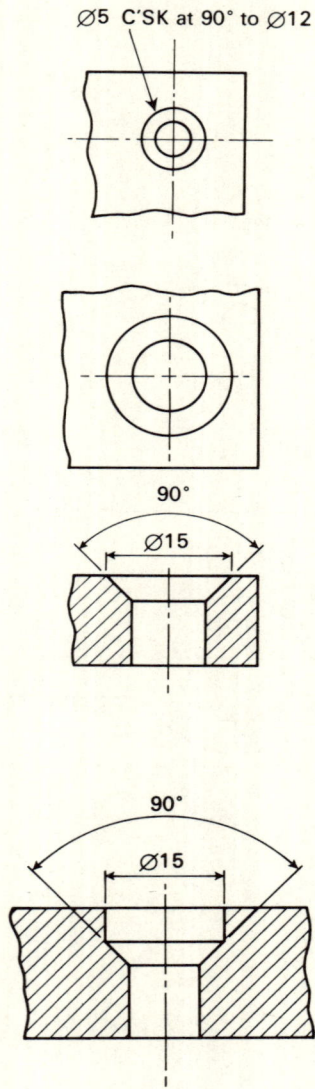

Figure 1.29 Countersinks

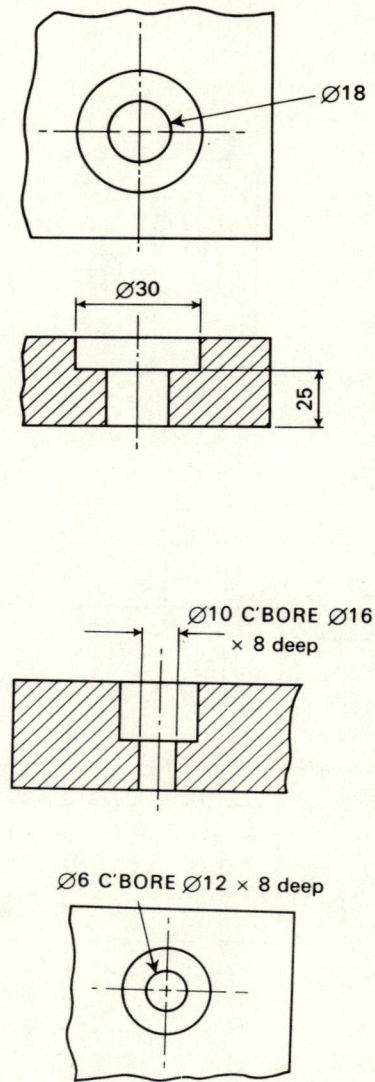

Ø18

Ø30

25

Ø10 C'BORE Ø16
× 8 deep

Ø6 C'BORE Ø12 × 8 deep

Figure 1.30 Counterbores

Ø18 S'FACE

Ø10
S'FACE Ø20

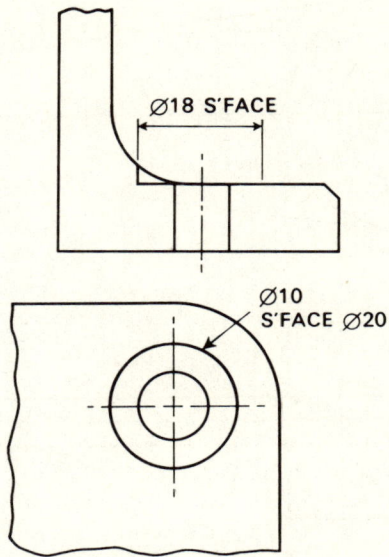

Figure 1.31   Spotfaces

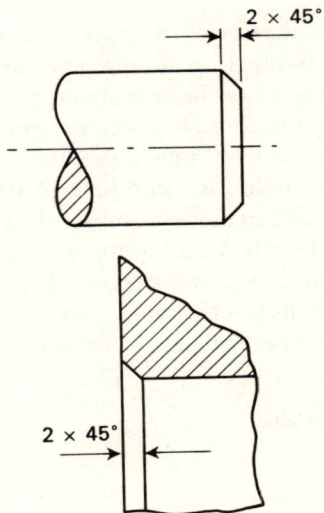

2 × 45°

2 × 45°

Figure 1.32   Chamfers

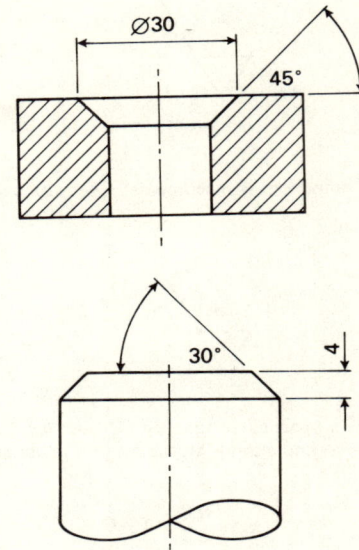

Ø30

45°

30°

4

Figure 1.32 *(continued)*

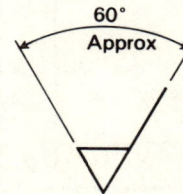

60°
Approx

Figure 1.33   Form of symbol—indication that a surface is to be machined, without stating either the surface texture grade or particular process to be used

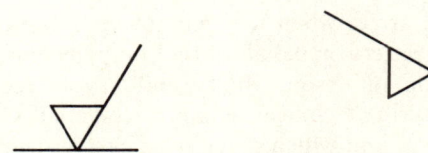

Figure 1.34   Application of symbol normal to the line representing the surface

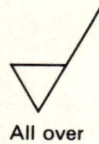

All over

Figure 1.35   Where a component is machined all over, use a general note

4.2 μm

Figure 1.36   Indicating a particular quality of surface texture expressed numerically in micrometres, and stating maximum permissible roughness

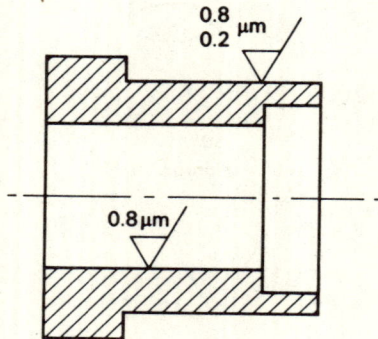

0.8 μm
0.2

0.8 μm

Figure 1.37   Indicating maximum and minimum values of roughness

The time will arrive when it is part of your responsibility to dictate and sign letters on behalf of the company employing you. The combination of design ability and a good command of language to describe technical details is essential. Consider the following information, which describes a cast lever (figure 1.38).

The lever consists of a cylinder 40 mm long and 50 mm in diameter with a concentric hole of 30 mm drilled through it. The

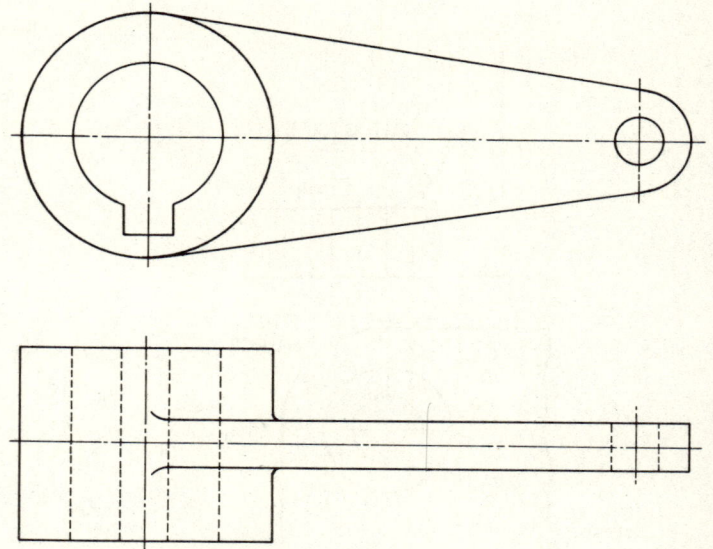

Figure 1.38

arm of the lever has its principal axis perpendicular to the axis of the cylinder and extending from a point halfway along the length of the cylinder. The sides of the arm are tangential to the curved surface of the cylinder and also to the circular end of the arm. The circular end has a radius of 10 mm. Its centre is 110 mm from the central axis of the cylinder. A small hole of 10 mm diameter is drilled concentric with, and passing through, this circular end. The arm is a constant 10 mm thick throughout its length. A rectangular keyway of 10 mm width and 5 mm depth is machined for the whole length of the large hole. The two sides of the keyway are perpendicular to the principal axis of the arm.

Methods of Dimensioning

Decimal Marker

The decimal marker for use with metric units is a clear point, which should be given a full letter space and should be placed on the baseline, for example, 42.25.

Units

The units of dimensions should be stated on the drawing and unless clearly inappropriate should be expressed in millimetres. The abbreviation may be omitted, provided a note similar to the following is used.

All dimensions are in millimetres.

When a dimension is less than unity it is recommended that the decimal sign be preceded by a zero, for example, 0.7.

Arrangement of Dimensions

Dimensions must be positioned so that they may be read either from the bottom or from the right-hand side of the drawing. The positioning of dimensions in the most suitable position on an engineering drawing is not necessarily an easy task. The detail drawing contains all the numerical information necessary for manufacture, and manufacturing errors due to poor positioning of dimensions leads to time and money spent on rectification.

A Reminder

The drawings you complete represent three-dimensional objects in two-dimensional form and convey geometric information. The drawings provide not just shape information but also give the sizes of features and state the accuracy to which they are to be manufactured.

CONVENTIONAL REPRESENTATION OF STANDARD PARTS

The student should refer to BS 308: 1972 for complete information on conventional representation of parts. Some typical conventions are provided here (see figures 1.39 to 1.43).

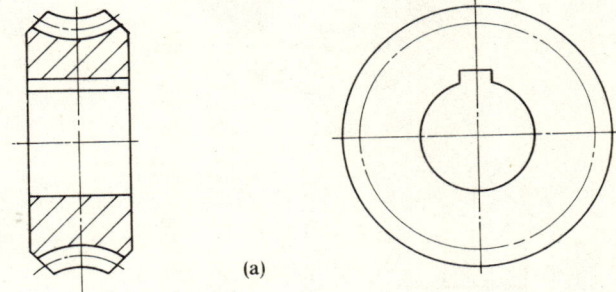

(a)

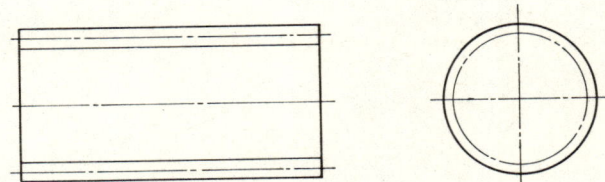

(b)

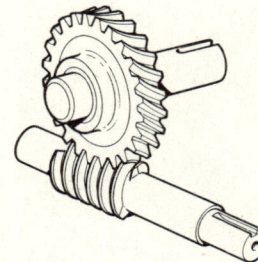

Figure 1.39 Conventional representations of (a) wormwheel and (b) worm

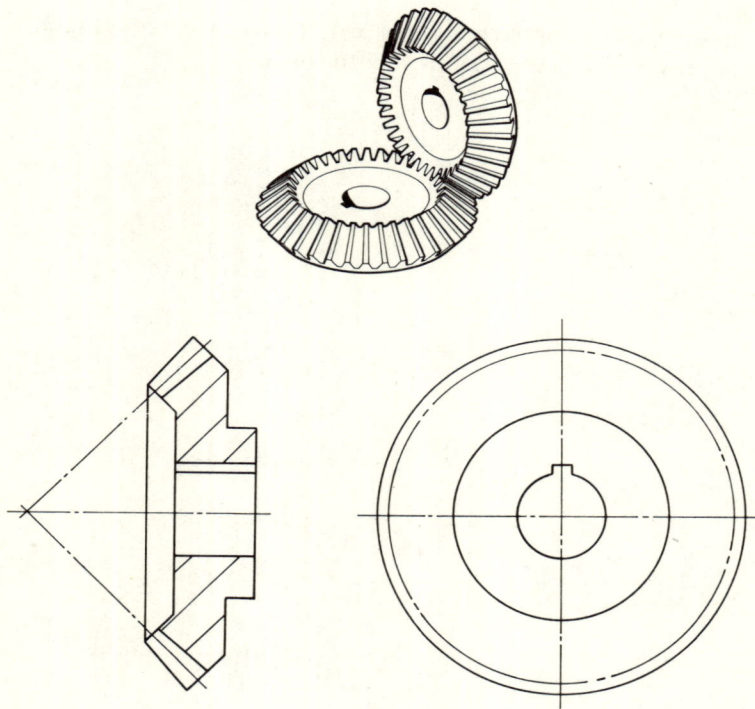

Figure 1.40   Conventional representation of a single bevel gear

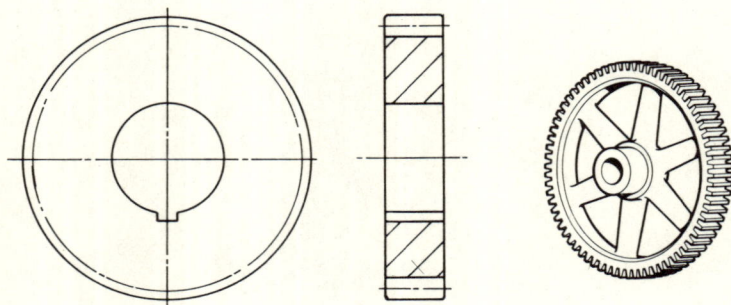

Figure 1.41   Conventional representation of a spur gear

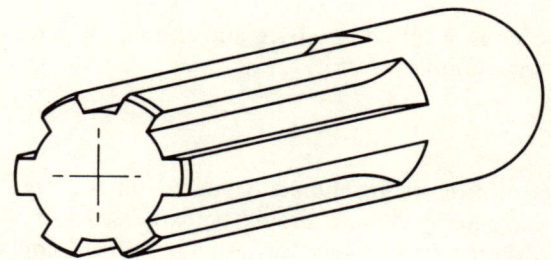

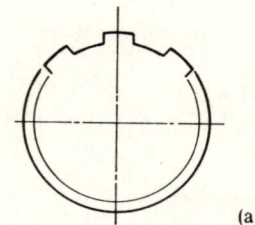

(a)

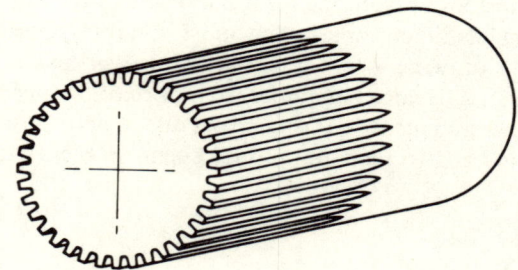

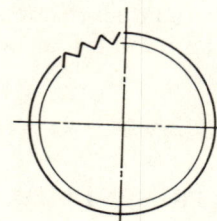

(b)

Figure 1.42   Conventional representation of (a) splined shaft and (b) serrated shaft

Outer race

Conical roller

Bore

Inner race

Roller cage omitted

(a)

Outer race

Cage

Inner race

Bore

Bore

Parallel roller

(b)

Bore

(c)

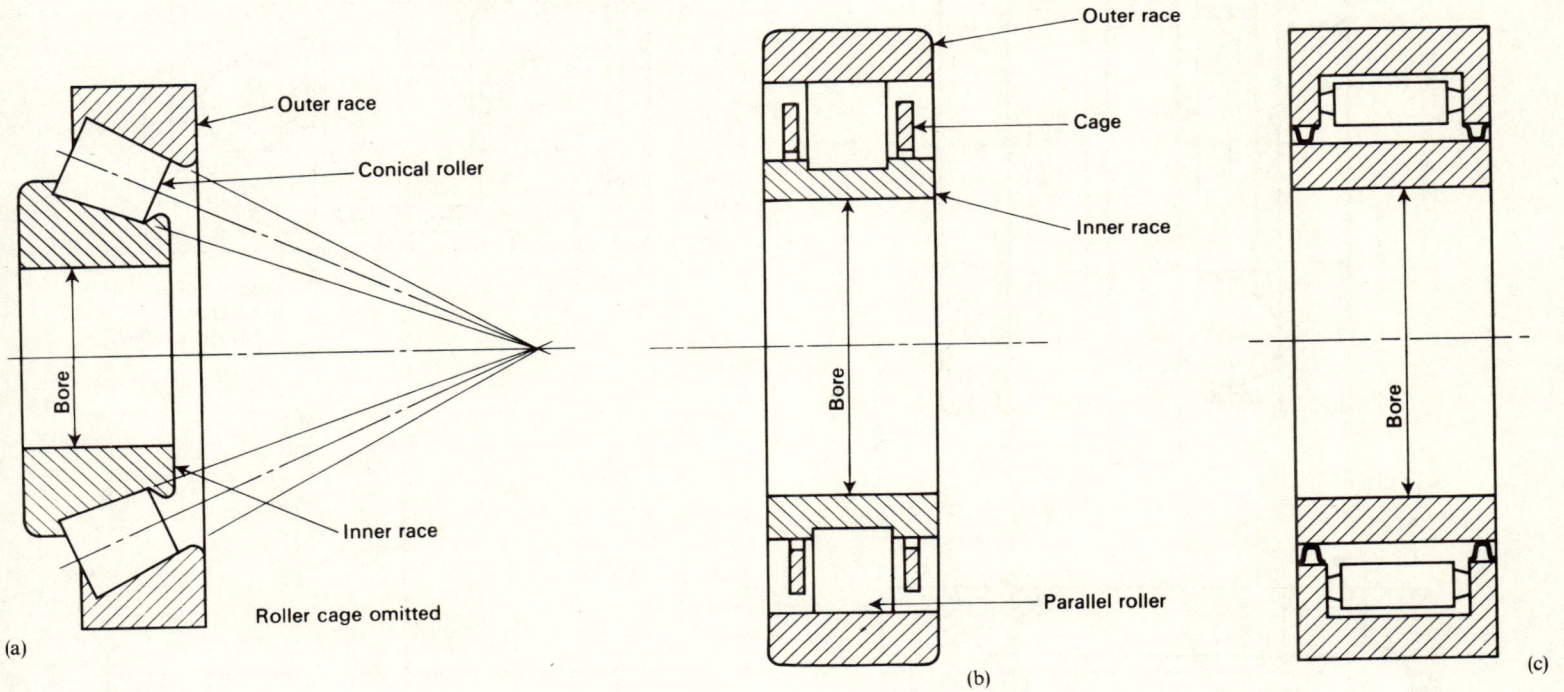

Figure 1.43 (a) Tapered roller bearing; (b) parallel roller bearing; (c) needle roller bearing;

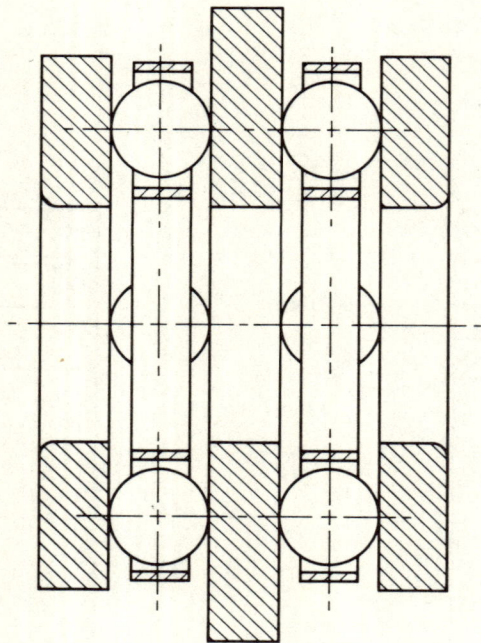

(d)

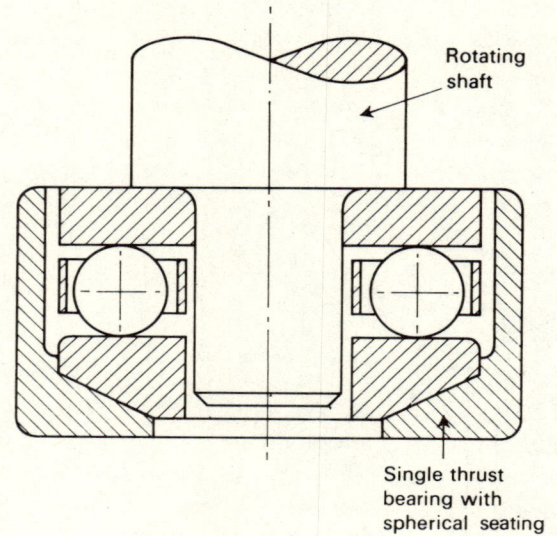

Rotating shaft

Single thrust bearing with spherical seating

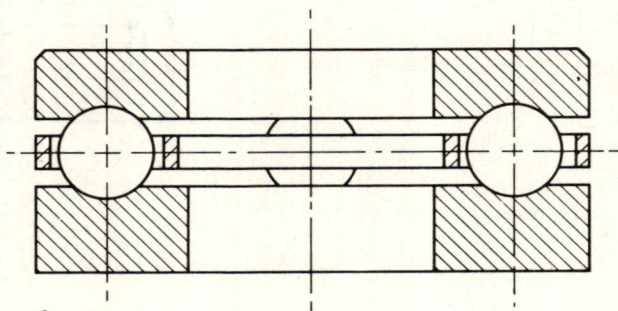

(f)

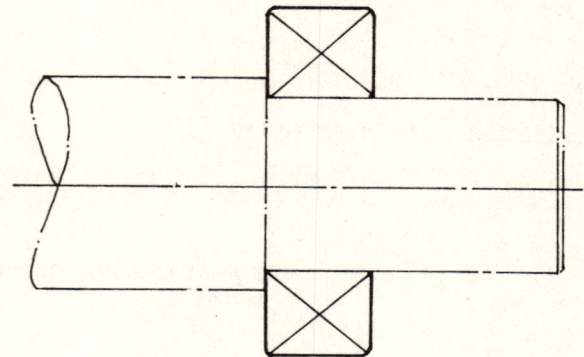

Figure 1.43 (*Continued*)   (d) double-thrust bearing; (e) single-thrust bearing with spherical seating; (f) single-thrust bearing; (g) conventional representation of a bearing

WELD SYMBOLS

A standard method of indicating weld symbols on drawings is essential, and the student should refer to BS 499: Part 2: 1965 which gives information as to the type, size and position of welds. The system is based on the following.

(1) A weld symbol
(2) A reference line
(3) An arrow
(4) A dimension

A weld symbol indicates the type of weld. A reference line is a datum line for a weld symbol whose position relative to this reference line locates the weld. An arrow indicates the reference or datum side of the joint, known as the arrow side. A dimension specifies the size of the weld required and is always placed to the left of the symbol (see figures 1.44 to 1.46).

Arrow side of joint

Arrow side of joint

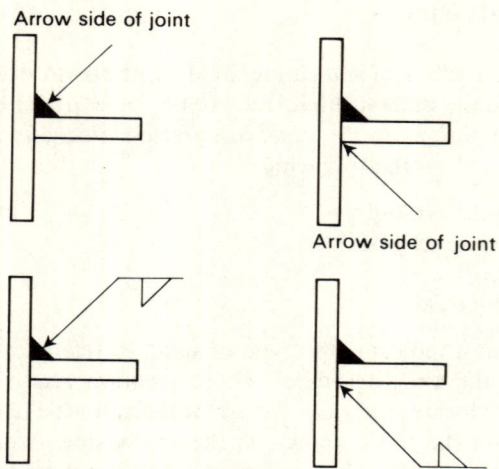

When arrow is on weld side of joint the weld symbol is INVERTED and suspended from reference line

When arrow is not on weld side of joint weld symbol is placed on reference line

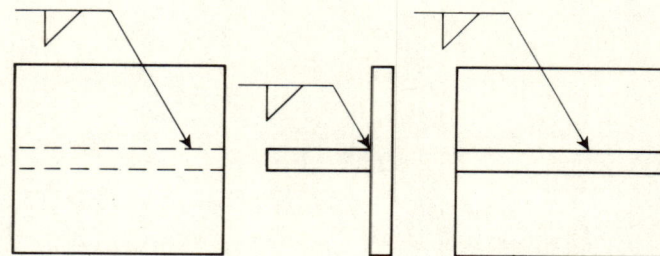

Arrow side NOT weld side

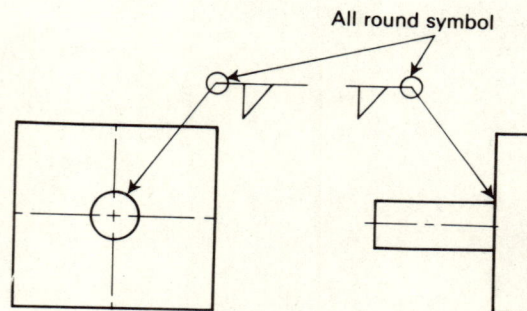

Arrow side is weld side

Note
In fillet welds the vertical leg of the symbol is always on the left hand side

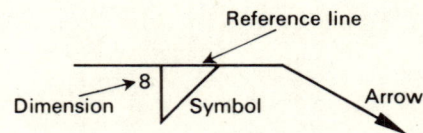

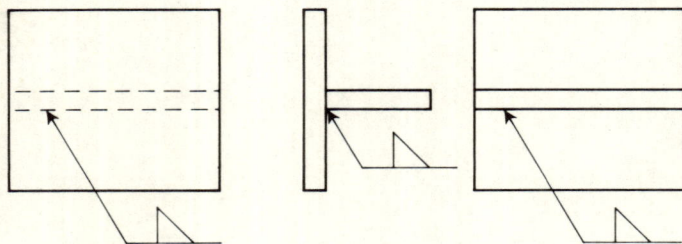

Reference line

Dimension   8   Symbol   Arrow

All round symbol

Figure 1.44 'Weld all round' symbol: When a continuous weld is required all round a component, an additional circular symbol is included as shown. In practice each weld is specified once on a drawing and shown on the most convenient view

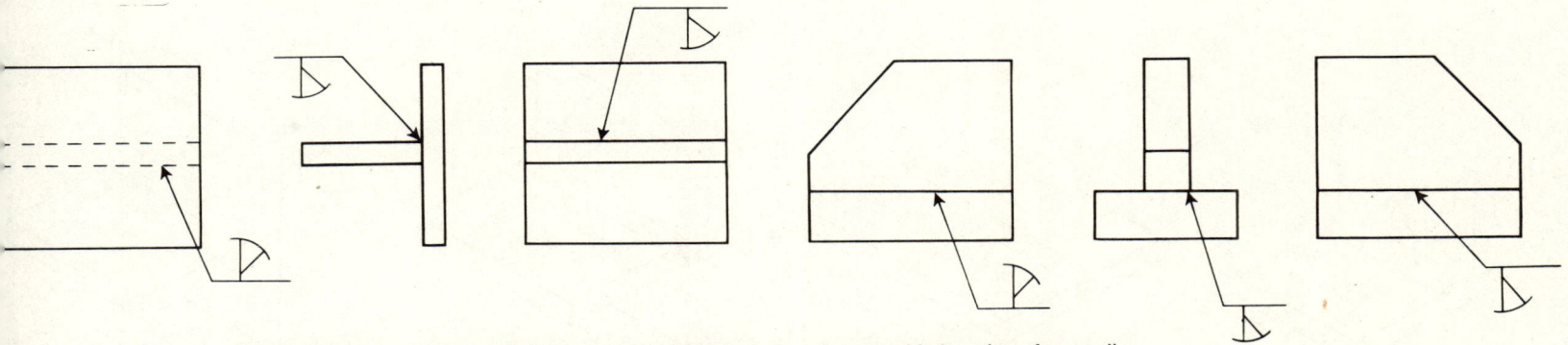

When both sides of the joint are to be welded the symbol is shown above and below the reference line

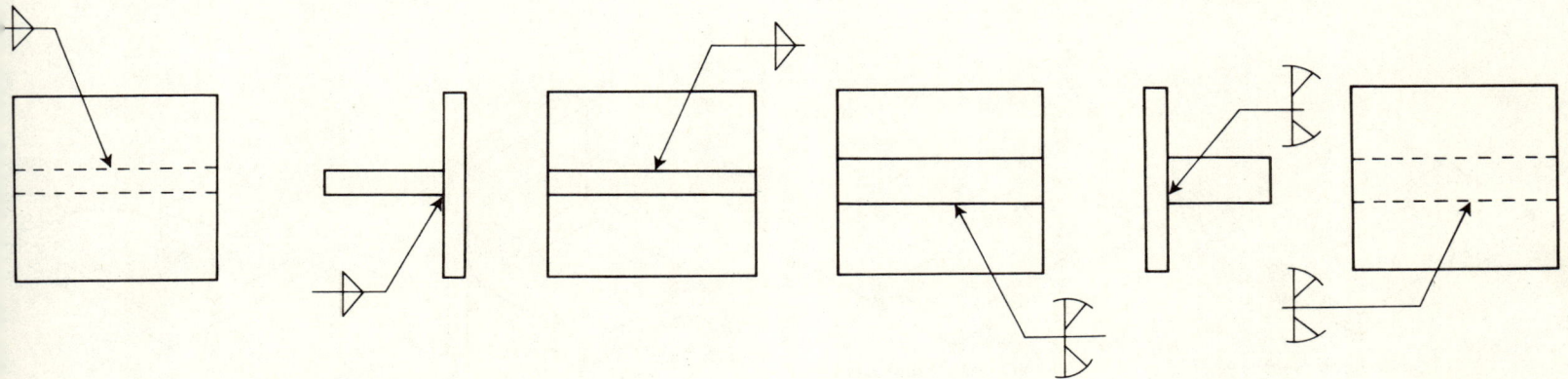

Note
In butt bevel welds the vertical leg of the symbol is always on the left hand side

Figure 1.44 *(continued)*

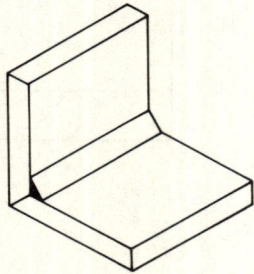

(a) Fillet weld

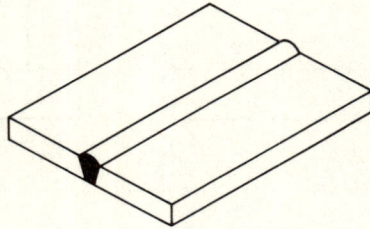

(c) Single V butt weld

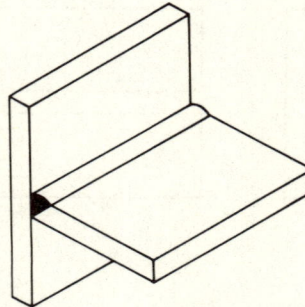

(e) Single bevel butt weld

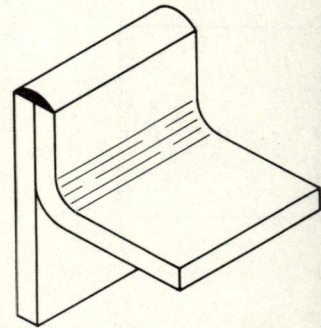

(g) Edge weld

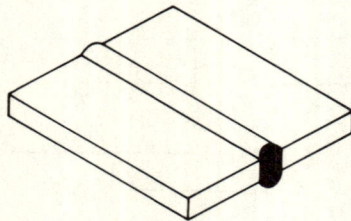

(b) Square butt weld

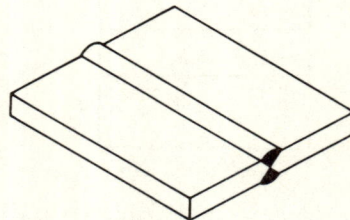

(d) Double V butt weld

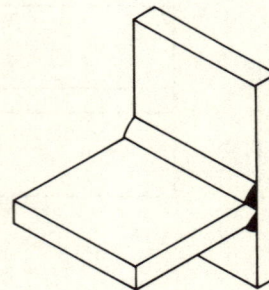

(f) Double bevel butt weld

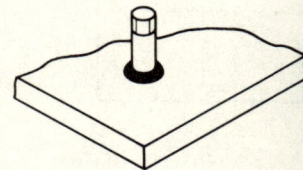

(h) Stud weld

Figure 1.45

SKETCH OF JOINT	TYPE OF JOINT	TYPE OF WELD	PLATE THICKNESS	PLATE EDGE PREPARATION	SYMBOL
	BUTT	BUTT	UP TO 3mm	SQUARE	⊤⊤
	SINGLE VEE BUTT	BUTT	OVER 5mm	BEVEL	◇
	DOUBLE VEE BUTT	BUTT	OVER 12mm	DOUBLE BEVEL	⊗
	LAP	FILLET	OVER 5mm	SQUARE	▷
	CORNER	FILLET	OVER 5mm	SQUARE	△
	CORNER	FILLET	OVER 5mm	SQUARE	◁
	TEE	BUTT	OVER 12mm	DOUBLE BEVEL	⫶

WELDED JOINTS AND SYMBOLS

Figure 1.46

SOME GENERAL PRINCIPLES

Illustrations showing some general principles of draughtsmanship are provided (see figures 1.47 to 1.58).

SIZE OF DRAWING SHEETS ISO 'A' SERIES	
DESIGNATION	DIMENSIONS mm
A0	1189 × 841
A1	841 × 594
A2	594 × 420
A3	420 × 297
A4	297 × 210
A5	210 × 148
A6	148 × 105

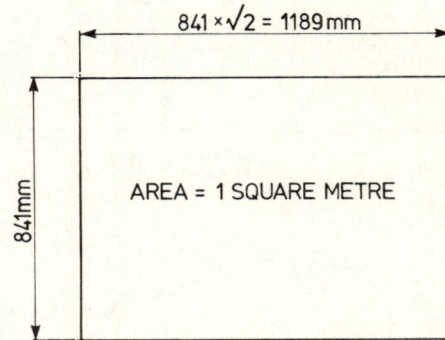

Figure 1.48   The standard basic size of drawing sheet is a rectangle of area 1 m^2 with sides in the ratio 2:1 as shown—this is known as A0 size. The other paper sizes are obtained by halving the preceding sheet along its longer edge. For drawing exercises the student will normally use A2 size

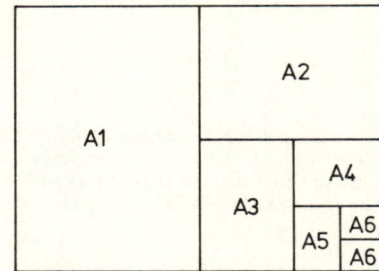

Figure 1.49

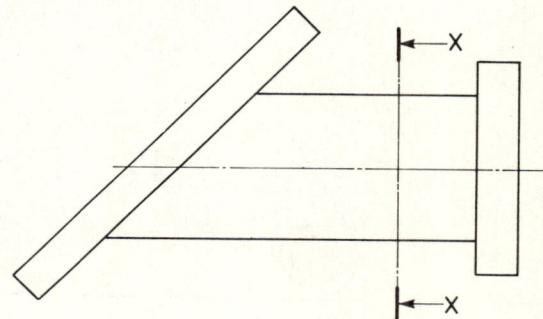

Figure 1.50   Cutting planes should be indicated by long chain lines, thickened at the ends and at a change of direction; they should be designated by capital letters. The direction of viewing is shown by arrows resting on the cutting line

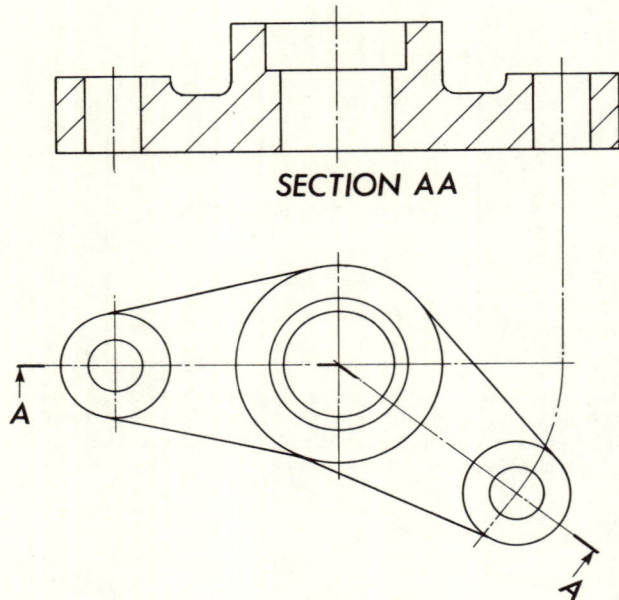

SECTION AA

Figure 1.51   A section taken along the horizontal centre of the component shown will provide information only for the left-hand arm feature. If the cutting plane is positioned to pass through both features and considered as a single plane, then the full shape description can be shown in two views. This is an *aligned* section

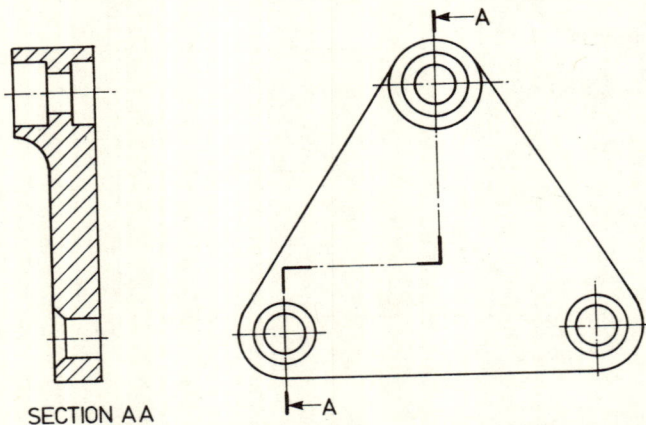

SECTION AA

Figure 1.52   The *offset* section is so called because the cutting plane is offset to include features that could not be shown in section if a single cutting plane were used

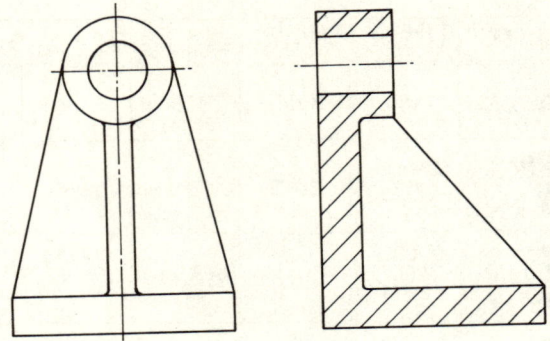

Figure 1.53   Ribs or webs are not shown sectioned (cross-hatched) when the cutting plane passes through them longitudinally; the same applies to bolts and shafts

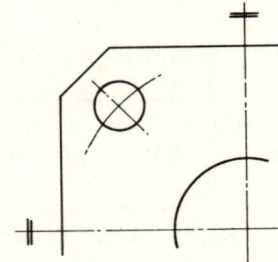

Figure 1.54   Sometimes it is necessary to draw symmetrical components in full. In such cases the line of symmetry is defined by two short thick parallel lines drawn at each end of, and at right-angles to, the symmetry demarcation line. Note how the outline of the component is extended slightly beyond the line of symmetry

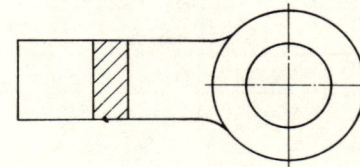

Figure 1.55   Revolved sections show the shape of the cross-section on the actual view of the part; the cutting plane is revolved in position

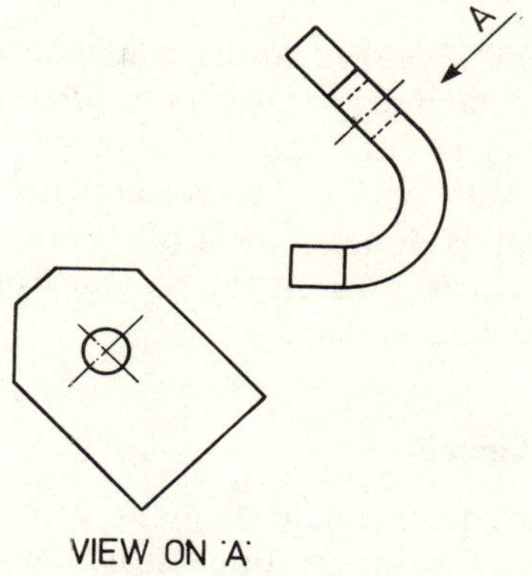

A

VIEW ON 'A'

Figure 1.56 Components with inclined faces can have such faces projected to show the true shape of the inclined surface; this is called an *auxiliary view*. The note and arrow may be omitted if the meaning is clear without them

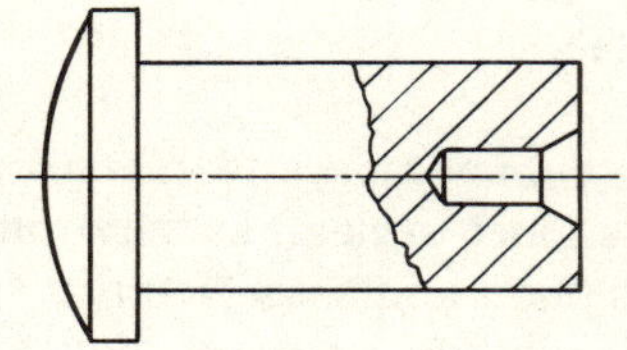

Figure 1.57 If it is not convenient to show a complete section or half-section of a component, a *local* section may be drawn. The local break is shown by a thin continuous irregular line

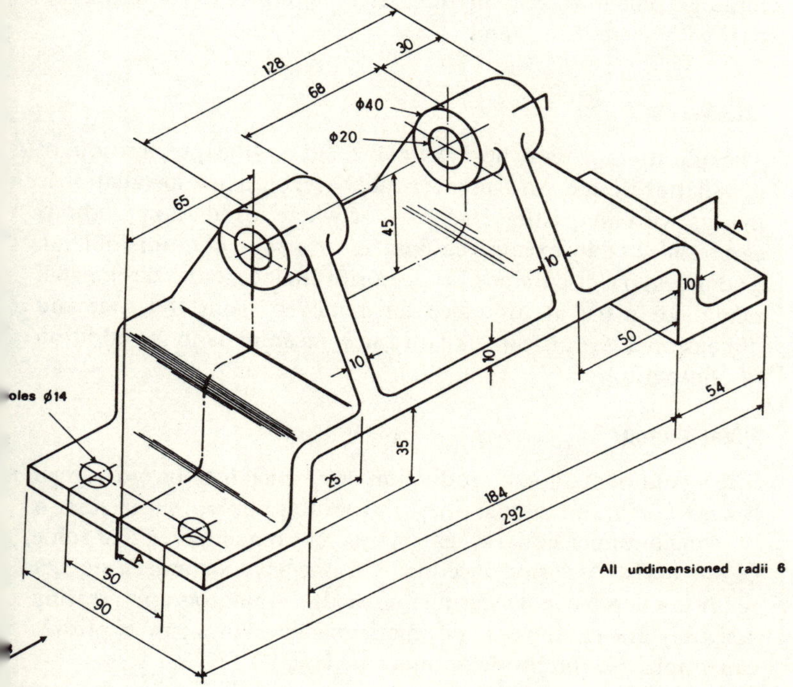

128
68
30
Φ40
Φ20
65
45
10
10
10
50
10
oles ø14
35
25
50
184
292
54
90

All undimensioned radii 6

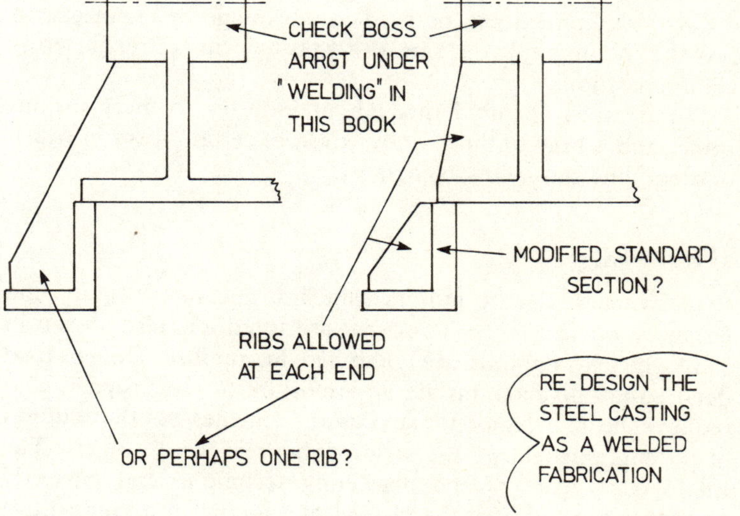

A

CHECK BOSS
ARRGT UNDER
"WELDING" IN
THIS BOOK

MODIFIED STANDARD
SECTION ?

RIBS ALLOWED
AT EACH END

OR PERHAPS ONE RIB?

RE-DESIGN THE
STEEL CASTING
AS A WELDED
FABRICATION

Figure 1.58 The pictorial view shows a steel casting. As a design exercise, prepare a fully dimensioned detail drawing of this component, in the form of a welded fabrication. Show a sectional elevation on AA and end view on arrow B. Dimensioning, arrowheads, layout and welding symbols to the relevant British Standards must be used. (Note the suggested design features)

PHOTOGRAPHY AS A MEANS OF ENGINEERING COMMUNICATION

Photographs are so commonplace that, generally speaking, they are accepted with as much acclaim as the ordinary nut and bolt. The definition that a photograph is a picture taken by means of the chemical action of light on sensitive film, tends somehow to minimise the great advances in film, camera and processing techniques that have taken place over the years. To say that 'every picture tells a story' is so very true; even the cheapest camera can, in the hands of a novice, capture a moment in time for all to see. There are many thousands of books describing the types of camera, the various developing processes and techniques, and the many other aspects of photography which are all part of this highly skilled profession.

The camera equipment for the production of catalogues, brochures and instructional manuals is sophisticated and so is the processing. However, there are other instances where quite ordinary equipment can be used to advantage by engineers, to provide photographs as an everyday means of engineering communication.

Photographs are used throughout industry to meet various needs, and a brief outline is now given of certain ways in which photography can help the engineer.

The Catalogue

Wall switches, electric motors, clutches, pneumatic equipment, fasteners, oil seals—the items that are listed and also shown in catalogue photographs are unlimited in number. Dimensional details and line diagrams are all-important to the designer, who requires the information for layouts and schemes, but the addition of photographs shows the style of the 'real-world' object. The catalogue is part of the marketing technique and correctly presented can enhance the chance of winning in a competitive market. Also, the photograph in a catalogue can present clearly, to perhaps a less technically minded person, a grasp of the problem in question.

The Brochure

The brochure is similar in style to the catalogue, but whereas the catalogue shows a complete range of products, the brochure usually shows either a single item or a range of similar items. Although there is sometimes less technical detail in the brochure format, the photographic details will often provide interest and information sufficient to warrant the request from a prospective customer for more detailed facts.

The Instruction Manual

Supplied with new equipment, as an aid to commissioning or for routine maintenance purposes, the instruction manual can use photographs to advantage. Although the manual may contain excellently worded instructions and diagrams, the sequence of assembly or the sequence of removal of parts can be seen clearly in photographic form, leaving the user in no doubt as to the way to deal with the task in hand.

Site Work

The photograph can be a valuable aid in the preparation of installation design. An example could be where a new installation is mating with an existing structure, or where installation methods can be affected by existing equipment or certain contours of land. The necessary discussions between site engineers and design staff may take place at an office far removed from the site, and dimensional layouts can take on a new meaning with the addition of photographs.

Plant Layout

The layout of plant in a production unit which is to be redesigned is, once again, an example where the photograph can supplement a drawing or dimensional scheme. It must be remembered that some of the staff at 'decision meetings' do not have, or indeed do not require, a depth of engineering knowledge. A photograph showing assembly line equipment, automatic machines or a row of presses can emphasise the problem under review.

Machine Details

Photographs of gearboxes, cross-slides, tooling arrangements and machining sequences can be of benefit to the design staff, the planning department and other sections of an industrial organisation. Although modifications to machines and equipment are recorded in drawing form, a photographic record can sometimes be beneficial (see figures 1.59 and 1.60).

Figure 1.59

Figure 1.60

Accidents

How was the workpiece supported in the machine? Were the press guards in position? What was the condition of the floor area? These and many more questions following an industrial accident can perhaps be partly answered by the use of on-the-scene photographs.

The foregoing are a few ideas concerning the use of photography as a means of engineering communication. There is also the testing of specimens, and the use of models, prototypes and finished products, which may require a photographic record. Presentation may be in cine-film form, or still form using, according to requirements, expensive or not so expensive cameras, or perhaps the instant developing film type of camera. Whatever the choice to meet a particular need, the photograph in colour or monochrome can present factual information of a real situation.

THE MODEL AS A MEANS OF ENGINEERING COMMUNICATION

A large design group can consist of designers, draughtsmen, clay modellers and craft fabricators in wood, metal, plastics and other materials. The design department considers all safety and ergonomic factors and parameters that concern the creation of, for instance, motor vehicles. The designer will base his preliminary sketches on his interpretation of the market research information available, together with his own sensitivity to the emergence of trends and requirements. From the sketch stage, full-scale layout drawings will be produced, and finally full-scale models. The full-scale models in clay are evaluated and are then cast and finalised as glassfibre models, which will be virtually indistinguishable from normal running vehicles. When production of a new car commences, millions of pounds will already have been invested in tooling requirements. It is therefore essential that all the problems within the design proposal are fully appreciated at the earliest possible stage, by the use of models. The model can be regarded as an excellent decision-making tool.

Models are valuable visual aids because of their power to communicate the same information to various persons engaged in different areas of work.

An instance of the use of models over 100 years ago is worth considering. Although their use was often frowned upon, the far-sighted engineers and scientists of the day were soon to prove the value of models in furthering new inventions.

Marine engineering in the early nineteenth century was under criticism with regard to science as applied to practical shipbuilding. In other words, knowledge regarding shapes of bow and stern and the theory of fluid resistance played little part in naval architecture. This period, of course, saw the introduction of iron as the material of construction, replacing wood which had been used for centuries. Also, the sail was being replaced by steam power.

At this time, John Scott Russell, a shipbuilder, studied the theory of fluid resistance on ships, experimenting with ships of different form, and having lengths of up to 25 m, towed by horses along Scottish canals. His experiments and research led him to the study of the shape of vessels, which meant the search for the ideal shape for a floating body that could move with the greatest ease through water. The British Association sponsored his work, and by the mid-nineteenth century he had made observations of tests in canals and in the sea with many vessels.

A few years later Charles Merrifield, Principal of the Royal School of Architecture, doubted that Russell had achieved the ideal shipform, and he suggested to the British Association that full-sized ships be used, towed at various speeds, to measure the resistance to motion and to study the direction and velocity of the accompanying water currents.

A civil engineer, William Froude, disagreed with the experiments proposed by Merrifield. Froude was convinced that to use small-scale models was a far better experimental technique to employing ships of full size. He argued that the cost of full-sized ships built purely for experimental purposes was prohibitive and that the results of trials with an actual ship were defective. For instance, was the measured resistance due to currents created by the wake of the towing vessel or perhaps by the wind? In the case of a steam-powered vessel, an inefficient engine could affect the ship's motion. Froude expressed his views to the Admiralty, asking them to

finance him in the building of an experimental tank near his home at Torquay. He proposed a covered waterway, 80 m long and protected from the weather. He recommended a model ship of about 2 m in length which would be towed by a steam-powered truck providing uniform motion. A dynamometer would record the resistance and the tracing of lines on a moving sheet of paper would be used to record the velocity of the ship, and at the same time would exclude personal observation errors.

Froude would have pleased the present-day designer by his materials/cost evaluation reasoned approach to model requirements. He required a material for his models which would allow cheap manufacture and quick modification due to the variables that had to be met in the experiment. The specification therefore required a material that was easy to cut, had a smooth finish and was waterproof. Froude decided on wax. It was fusible, it could be cast in moulds, with adjustments made by cutting, and it could be recycled by melting for further models.

The Admiralty approved Froude's scheme and his experiments were successful. His scientific methods and approach, indicating optimum speeds, proportion and forms, were to influence future ship design. One of his findings, for example, was that ships were being built that were not broad enough at midships in proportion to their length.

The brevity of this early example of modelling does not give full justice to the characters involved. However, the student should follow the approach to the problem made by Froude and match it against the systematic approach to design outlined elsewhere in this book. Under the heading of 'The Feasibility Study' in chapter 4, it is stated that consideration must be given to the economic, physical and financial aspects of the design. In asking the Admiralty for their support, Froude argued that even if the improved designs only reduced the power required for propulsion, savings would be made on the national coal bill. Also, if a ship could be more efficient by building it 3 m shorter, the materials saving would justify his experiments. Froude's models certainly proved to be excellent decision-making tools.

Let us consider the diagram which illustrates two types of model used for the same proposed building (see figure 1.61). Both models

MODELS

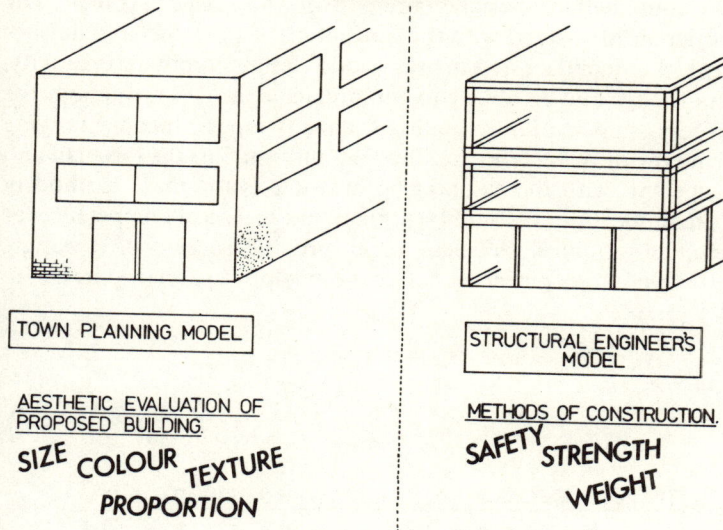

TOWN PLANNING MODEL

AESTHETIC EVALUATION OF PROPOSED BUILDING.

SIZE COLOUR TEXTURE PROPORTION

STRUCTURAL ENGINEER'S MODEL

METHODS OF CONSTRUCTION.

SAFETY STRENGTH WEIGHT

Figure 1.61

are incomplete in that features such as heating and ventilating systems, plumbing arrangements and internal wall finishes are not shown. In each case only certain properties of the finished building are shown, and this typifies an essential characteristic of all models. They are manufactured to serve a purpose and, in the case of the town planning model, the aesthetic features are of prime importance. To the construction company, the importance of the model is in the construction features that are shown. Both models, if true to scale, can give a guide to both interested parties. If there is disagreement over certain features, the models can be modified until agreement is reached.

In the field of engineering technology, the model is an essential part of the design function. It can be manufactured before the prototype and it is capable of being altered at small cost. If it gives a true representation of the prototype, then we can say that the model is a valid model. However, when should a model be used to help design decisions? Equally important, can we construct wooden, plastics, glassfibre or wax models for use in every type of design approach? Consider the following summary.

Economy of design can often be achieved by the use of accurate models. Although sophisticated models increase design costs, savings on the manufactured products can be achieved through their use. In the case of a mass-produced article, a small saving on each component part of an assembly can justify more time spent on design work. A similar type of reasoning applied to a large project could lead to large savings in materials.

Safety is emphasised in many ways throughout our daily life. The structural engineer will use a factor of safety according to the requirements of the structure in question. Any shortcomings in the model are balanced by his knowledge and use of factors of safety. The designer of an aircraft or spacecraft requires more sophisticated models to achieve an overall balance of safety, payload, flight distances, speeds and fuel consumed.

Designers often have to look for solutions where models in the form of Standards and Codes of Practice are used, which provide information that in itself is a guarantee against failure.

The necessity for this form of model is essential in cases where detailed modelling is impossible, owing to the complex behaviour of features or components. The welding of pressure vessels, complex arrangements of rivets and fasteners and underground tunnelling are examples. Research establishments often carry out tests using full-sized equipment to achieve the data and information provided in the various Codes of Practice.

Mention should be made of the combined use of the wind tunnel and the model. Although usually thought of in terms of aircraft research, the wind tunnel can be used for the 'drag' measurements on cars and the trajectory analysis of missiles or bombs released at speed into steady winds. The modern oil rig towering hundreds of feet above the sea, with a platform the size of a soccer pitch, can require tests involving the wind tunnel and a model, by which winds of differing velocities and turbulence gradients can be simulated.

The features of stability and function can be used and tested to good effect using timber, plastic, cardboard, plasticine and construction kits of the Meccano type. It is hoped that, as students, you will further your knowledge of design studies by constructing models using a logical approach.

2 The Engineering Designer

We could define a designer as a person who solves problems. The design process starts with the conception of a product and finishes with the manufactured article. Good design combines creativity, knowledge and an ability to communicate ideas. Throughout the process, information is imparted to many types of people, ranging from the drawing office and buying office staff to the foreman and inspector, with the engineering drawing as the main method of communication. It could be argued that technical competence, or even intuition, is sufficient to ensure the success of a design. However, the engineering designer can adopt a special approach to

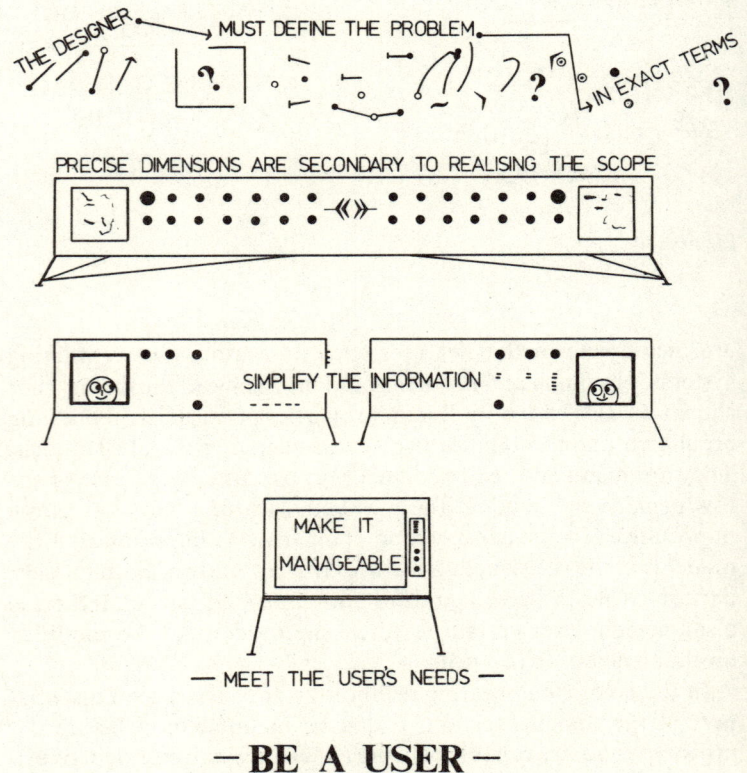

THE DESIGNER → MUST DEFINE THE PROBLEM IN EXACT TERMS ?

PRECISE DIMENSIONS ARE SECONDARY TO REALISING THE SCOPE

SIMPLIFY THE INFORMATION

MAKE IT MANAGEABLE

— MEET THE USER'S NEEDS —

— BE A USER —

Figure 2.1

professional problem-solving by combining technical competence with a systematic approach to all aspects of design.

Once presented with a problem the designer must then start to define it in exact terms (see figure 2.1). Having done this, a systematic design approach can be adopted, not forgetting that the designer can, during the approach, return to an earlier stage and make fresh decisions (see figure 2.2).

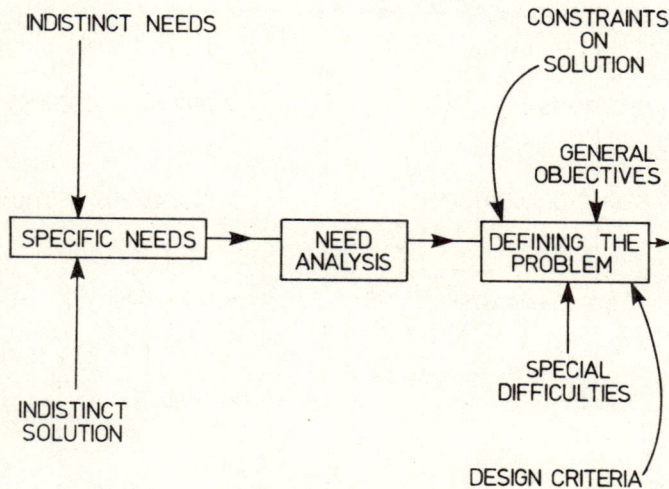

Figure 2.2

Terminology can differ. For instance, we may use the term 'specific need' or 'user's need' in place of 'problem'. The need must be analysed and then followed by an exact definition of the problem. The job of the designer is to meet the user's needs, no more and no less, and the designer must be given a degree of freedom.

Certain basic essential information should be available to the designer at the commencement of a project.

(1) Information to evaluate the physical properties of the design

(2) General engineering and scientific information

(3) Product data

(4) Service information regarding the past performance of similar designs

However, there can be further considerations for the designer, without his even moving from the drawing board. Ideally, the task can involve the communicating of design solutions to those who are responsible for manufacture, and then to follow through the various stages of production intimately. This does not always apply, owing perhaps to pressure of work, size or type of organisation, or size of project.

The diagram provided shows the drawing as the link between the functions of design and production (see figure 2.3). The designer imparts requirements to the drawing office staff, who will be responsible for preparing and issuing drawings for the manufacture of the product. Distribution of facts entails close liaison as the communication system begins to enlarge. To stress this point, consider the position of the designer relative to the personnel involved in interpreting the designer's schemes into manufacturing

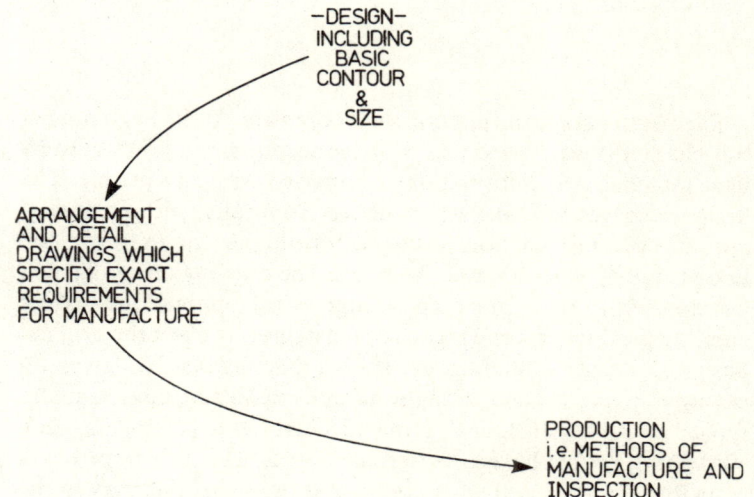

Figure 2.3

drawings. Such drawings are the universal language of communication for manufacture (see figure 2.4).

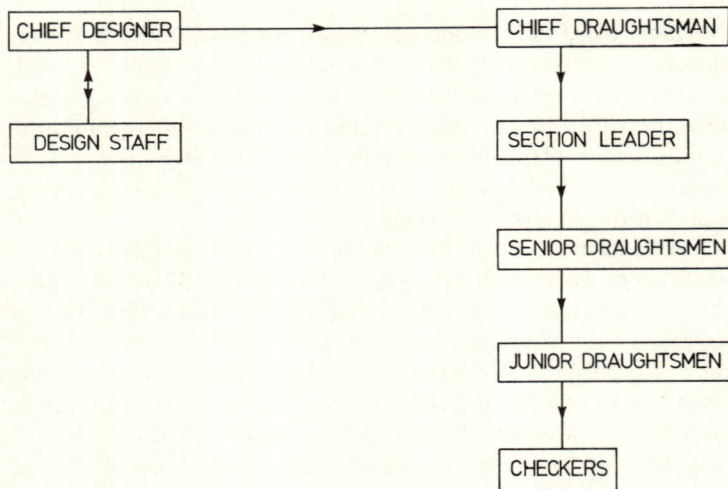

Figure 2.4

Even at this stage the picture is not complete. At the beginning of this chapter it was stated that, from conception of a product to its final manufacture, information is imparted to many people. The designer strives to make a product economically, without loss of appearance and function, using to advantage, for example, the knowledge of materials and processes. The drawing office staff will be expected to use similar knowledge as they pursue the task of creating production drawings that will in no way affect the original design concept. The diagram shown (see figure 2.5) gives an example of the design, draughting and manufacturing structure typical of an organisation. Note the various departments, each containing people having virtues and faults, who, it is hoped, will contribute to a success story. Industrial organisations vary in the type of technical structure adopted. There may be both a chief designer and a chief draughtsman. Some companies may not have

tracing facilities. However, the outline shown in the illustration will give some idea of the communication process that is necessary.

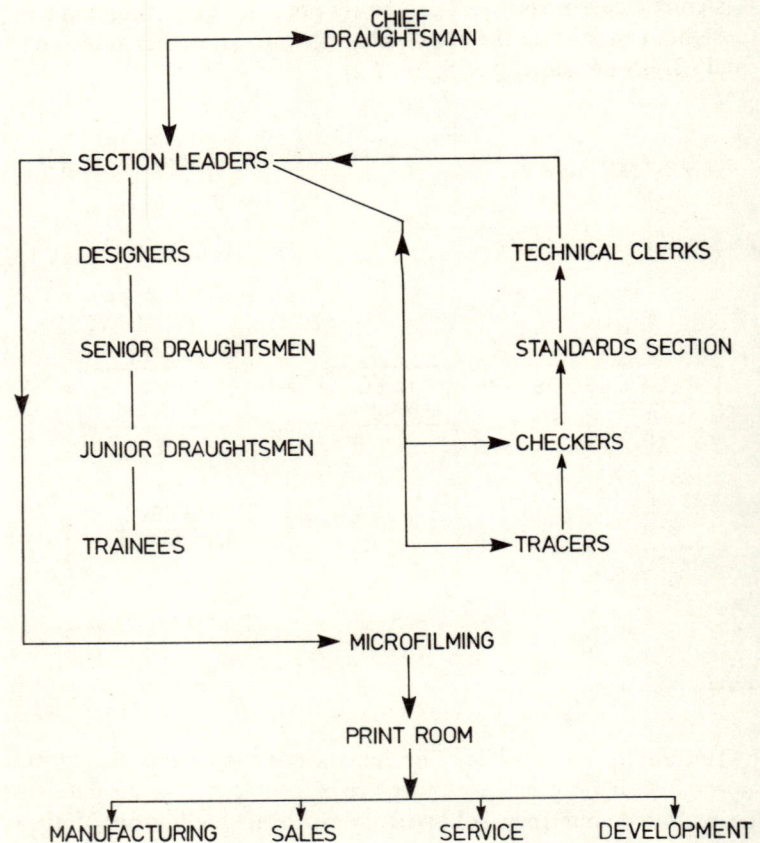

Figure 2.5

The Various Duties

The overall control of the drawing office is the function of the chief draughtsman. He will also interview and select staff for the office, and like any other leader he must have efficient staff within the department. Delegation of responsibility is not just a matter of getting rid of work in order to sit back and relax. Delegation is an art in itself, which can only be achieved correctly with full knowledge of facts and people. Too much delegation of responsibility is as bad as too little delegation—a compromise must be reached for effective decision-making. *Middle and top management must be technically competent as well as being controllers of labour.*

The work for the various sections is examined and distributed to the section leaders, who in turn are responsible for a team of draughtsmen. Upon completion of the work the section leader will submit it to the checking section. Certain organisations will then have the pencil drawings traced in ink on linen, thus providing a more permanent and durable record. The function of the standards section is to ensure that the standards information is in line with requirements. Parts lists and schedules, separate from those on the drawing, together with other technical paperwork, is the task of the technical clerks. All work of major importance is then returned via the section leaders to the chief draughtsman. Dyeline prints of drawings are produced and issued to the manufacturing units within the organisation, or perhaps to outside suppliers or contractors. Finally, the drawings may be microfilmed, which in both large and small organisations has advantages.

It is recommended that drawing sheets according to the International Standards Organization (ISO) 'A' series, and specified in BS 3429: 1961 be used (see table 2.1). A border of at least

Table 2.1

Designation	Overall Size (mm)
A0	841 × 1189
A1	594 × 841
A2	420 × 594
A3	297 × 420
A4	210 × 297

15 mm should be provided in the form of a frame to enclose the drawing area. It is important for a suitable filing system to be installed in the drawing office. Vertical filing cabinets or plan chest systems are the usual form employed.

Microfilming depends on a photographic process, and it is essential that a high contrast exists between the finished lines, dimensions, etc., and the paper background. The use of 35 mm roll film using an aperture card takes up only about 5 per cent of the space required by the original drawing. Used in conjuction with a viewer, the draughtsman can readily refer to the necessary requirements without constant handling of original drawings. Handling can cause deterioration, drawings can be misfiled and the use of microfilm allows reference desks to be used for essential drawing only. Drawings are valuable records and microfilming affords protection against damage by fire and theft.

PRELUDE TO DESIGN

All of us, at one time or another, have believed that certain articles or fitments could have been designed in a much better manner. There is always the inaccessible or irritating feature that can cause tempers to be frayed in the kitchen, the garage or the machine shop. It is easy to be wise after any event, and the user of a product can often find fault where the designer failed to do so. Designing, therefore, must not be an activity removed from real life. The designer must act in the interests of the purchaser or user. In this situation the complete picture is exposed instead of a basic framework, which is the hallmark of an armchair activity. At the commencement of design thought the precise dimensions are secondary to realising the scope and reality of a problem, and having collected what can amount to be a large quantity of information the skill of the designer is to simplify this information in order to make it manageable.

The designer may be part of a large design team involved with a complex machine tool, an aircraft or a car, with each designer having to convert thoughts into a limited space. The person who designs an instrument for a vehicle is not to blame for the poor positioning of that instrument on the vehicle dashboard. The persons involved are experts in their own particular field, each having to simplify the information to make it manageable. It follows that a large design project requires a large design team, which, by virtue of numbers, presents organisation problems.

The secondary feature of precise dimensions, mentioned in chapter 1, soon takes on an important meaning when the chief designer calls a meeting as things begin to take shape. The ample leg room, reclining seats, luggage space, adjustable steering column and panoramic style dash board will soon require a body the size of a juggernaut transporter. Many can say what has to be done and many can say why it has to be done, but only the good designer knows how the task is to be done. The good designer must have the ability to initiate, evaluate and interpret.

The designer is only one part of a system that creates man-made objects for everyday use. A simple cycle of events can be applied to most man-made objects. Most products have a life cycle (see figure 2.6). Once created, they progress into commercial use, where improvements are often made within the existing basic design. As time passes, new ideas become viable and the old product falls into disuse.

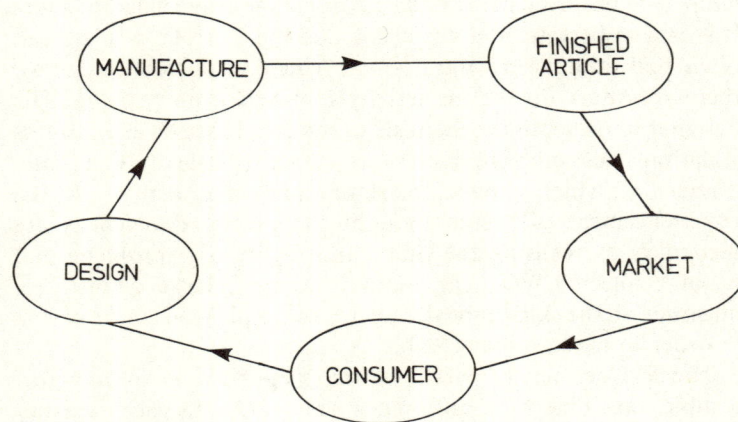

Figure 2.6

Designers and Designers

There are those who design wall-switches, power plugs, bottles, cutlery, chairs, telephones and spacecraft. Others are concerned with the design of systems or mechanisms to aid the production of these items, in the form of machines and attachments, which in turn may utilise electrical, hydraulic or pneumatic operational control. Whatever the task of the designer, the ultimate aim should be originality (if necessary), functional efficiency, economy of production and good appearance in the finished scheme. One may argue that whereas function and appearance are an essential relationship for cutlery or telephone design, the main consideration in the design of a mechanism to aid production of these items is that of function. However, just as the designer of a domestic appliance must consider the ease of movement involved to manipulate the appliance by the housewife, similar consideration must be given to a mechanism which is to be used by an operator in a factory. Both are operators of mechanisms and both are using energy. Such considerations regarding the physiological and psychological reactions of the operators to the environment in which they work are a specialist study. The reactions and behaviour of the operator and the stresses which he experiences, come within the field of ergonomics.

Let us imagine that a new product has been designed and that production is about to commence. The product, perhaps, is such that it incorporates all the good features of comfort and ease of control, affording greater cleanliness and an efficient utilisation of space. Whatever it is or whenever it is to be used, as it arrives on a certain section of conveyor track, within the production line, the design problem is to divert it off the main track on to a lower level conveyor for further treatment. The criteria of ease of control and utilisation of space, as applied to the design of the product, should also apply to the design of the production aid.

Consider now another diagram, showing a chain of thought used by the designer (see figure 2.7). In the previous consideration of the systematic approach to designing, you will have noticed that mention is made of 'constraints on the solution' (figure 2.2). These constraints can, for instance, be the lack of background knowledge of the designer, or they may be physical constraints, such as space

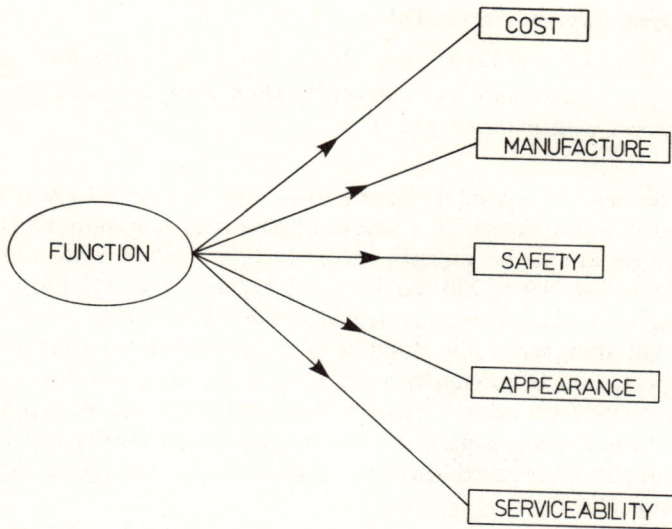

Figure 2.7

Method

Suggest reciprocating pusher rod, using basic pneumatic equipment, providing a fully automatic cycle (see figure 2.8).

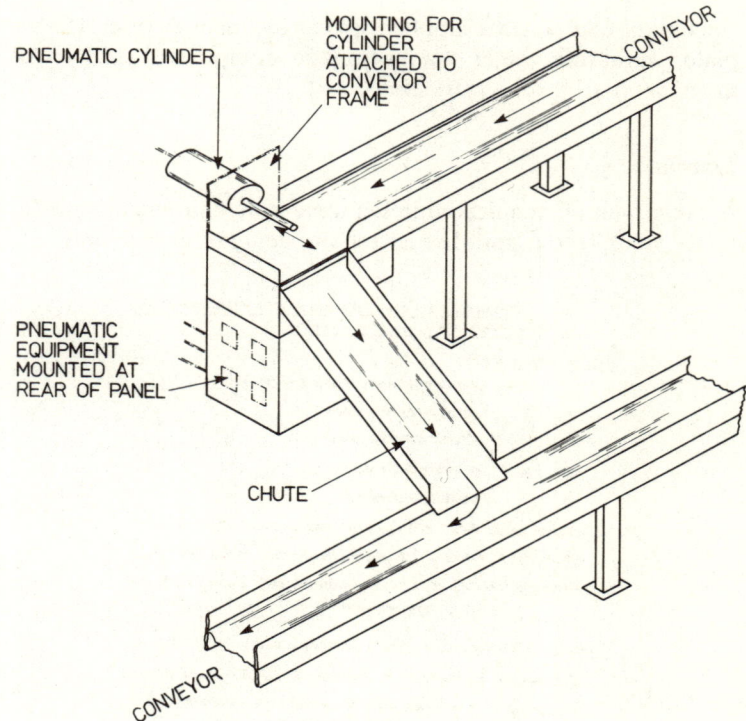

Figure 2.8

restriction. There is another constraint that can be applied by those who are in a position of control and feel that they know better than the designer. Such interference often results in a 'penny wise pound foolish' chain of thought. At its worst the chain of thought is function→ manufacture, which if enforced can belittle the designer, curtail ingenuity and professional integrity, and ultimately only highlights poor leadership.

Let us now consider the problem of handling our new product with the following approach.

Requirements

A means of diverting the product at 90° from the conveyor belt to a lower level conveyor. Manual performance was considered, but deemed much too tedious and labour cost prohibitive.

Cost

Air supply already available and in use on manufacturing line. Standard pneumatic components readily available.

Safety

All parts of the system will be suitably screened (arrange meeting with Safety Officer).

General

Pneumatic components supplied direct from manufacturer. Pusher plate, mounting panel for pneumatic equipment, etc., to be manufactured in own workshops.

Liaison

Arrange visit of technical representative from equipment supplier to see track layout and discuss requirements of pusher unit.

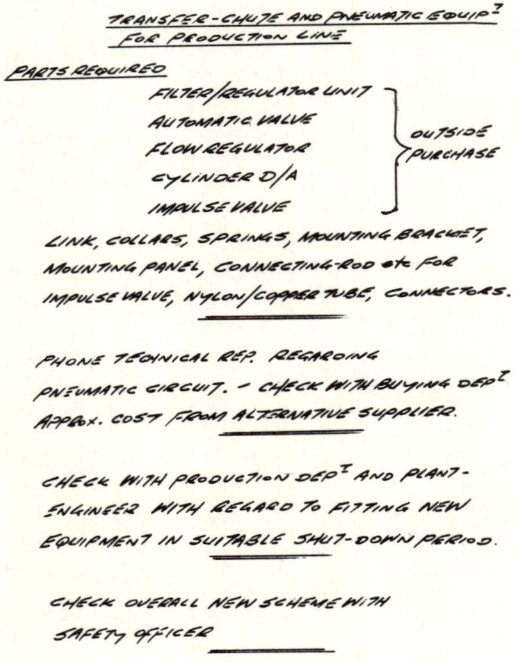

TRANSFER-CHUTE AND PNEUMATIC EQUIP.^T
FOR PRODUCTION LINE

PARTS REQUIRED
FILTER/REGULATOR UNIT
AUTOMATIC VALVE
FLOW REGULATOR   OUTSIDE
CYLINDER D/A   PURCHASE
IMPULSE VALVE

LINK, COLLARS, SPRINGS, MOUNTING BRACKET,
MOUNTING PANEL, CONNECTING-ROD etc FOR
IMPULSE VALVE, NYLON/COPPER TUBE, CONNECTORS.

PHONE TECHNICAL REP. REGARDING
PNEUMATIC CIRCUIT. - CHECK WITH BUYING DEP.^T
APPROX. COST FROM ALTERNATIVE SUPPLIER.

CHECK WITH PRODUCTION DEP.^T AND PLANT-
ENGINEER WITH REGARD TO FITTING NEW
EQUIPMENT IN SUITABLE SHUT-DOWN PERIOD.

CHECK OVERALL NEW SCHEME WITH
SAFETY OFFICER

Figure 2.9

Appearance and Serviceability

Suggest that unit is bolted to existing conveyor section for ease of fitment/removal. Check position of airline relative to gangway, etc. (discuss with plant engineer).

Remember, a good designer knows 'how the task is to be done', and from the outset has envisaged pneumatic operation. Existing airlines are to be utilised, incorporating the following typical details (see figure 2.9). Background experience and a thorough diagnosis of requirements help to make the decision.

The designer is not afraid or too high-handed to consult the experts, and as previously mentioned makes arrangements to put proposals forward to a technical representative from the suppliers of the pneumatic equipment. The manufacturers catologue will no doubt have been used to arrive at the initial scheme stage (see figure 2.10).

The technical representative, having approved the scheme, will submit a circuit diagram, together with a cost quotation. The circuit diagram illustrated (see figure 2.10) provides the basic information relating to pneumatic components and connections. However, the manufacturer will in addition also supply a similar diagram, using the British Standard symbols for the pnuematic components. By using standard symbols, a universal language of pneumatic circuitry is thus available. The student will, in his further design studies, be made aware of such standards (see figure 2.11).

The designer will, perhaps, at this stage require consultation with the company financial director or a similar member of staff, in order to obtain capital sanction for the equipment to be purchased. Once cost approval is obtained an order is placed, not forgetting that alternative quotations are almost always required (and necessary) in order to compare cost, quality and delivery from other suppliers.

The photographs (see figures 2.12 to 2.14). show equipment manufactured by Midland Pneumatic Ltd, Wednesfield, Wolverhampton. Such equipment is designed as an aid to production for many and various commodities. The piston movement can instigate mechanism motion in its many forms, and in design terms as the manufacturer of this particular equipment states: 'The wide experience of yesterday is built into equipment for tomorrow'. In terms of quality and reliability the equipment is robust, neat in appearance, manufactured from the correct materials and built to a high standard of workmanship. From the user's point of view it is therefore reliable and has a long working life. Such claims are also put forward by the designers and manufacturers of cars, washing

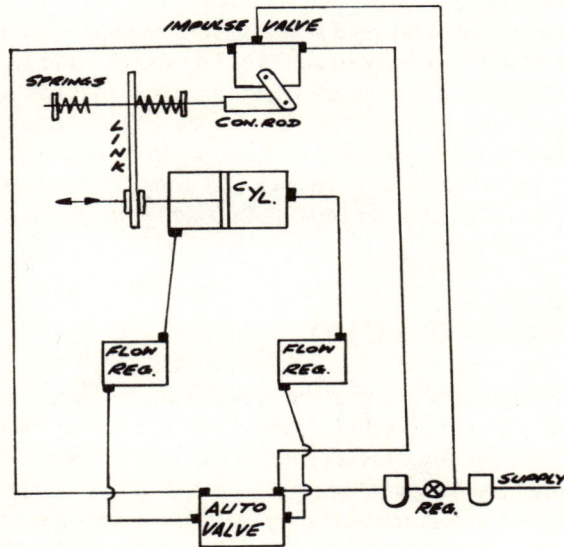

ROUGH SCHEME OF PNEUMATIC CIRCUIT
FOR RECIPROCATING UNIT

Figure 2.10 *Operational sequence*: Air supply from compressor unit passes through filter and pressure-regulator valve to supply port on automatic valve. This supply line is also connected to port on impulse valve. Automatic valve allows air to flow to cylinder via regulator. Piston rod with link moves forwards together with impulse valve connecting-rod. Length of piston stroke can be varied according to position of collars on connecting-rod. Short springs are fitted in front of these collars to compensate for very slight overrun caused by time lag between movement of impulse valve and response of cylinder. At completion of forward stroke, air flow is reversed via connecting link on impulse valve. Air flows into automatic valve and so into opposite end of cylinder. A continuous reciprocating cycle is thus available

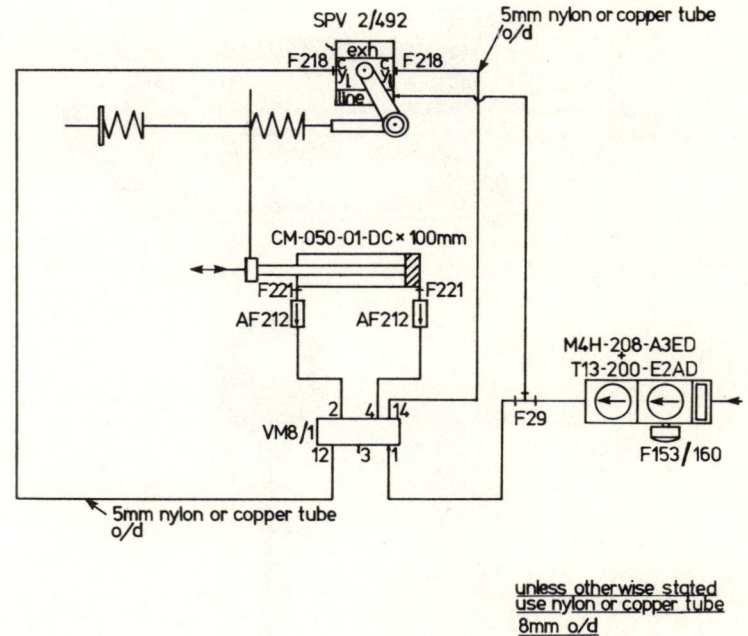

Figure 2.11

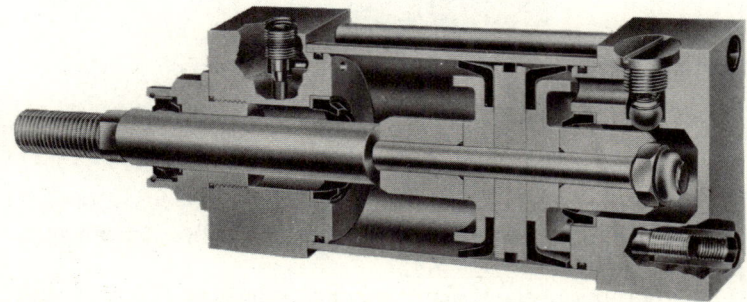

Figure 2.12

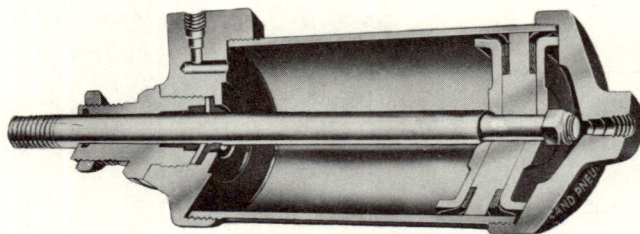

Figure 2.13

Figure 2.14

approaches, which within an industrial organisation can make life very difficult. Textbooks often assume that the toolroom, buying department and design office are similar in structure to a class at school, where registers are marked and feedback from questions occurs within a certain daily period. However, the chief buyer likes his say as much as the chief draughtsman, and the technical director is often not technical, and the steel supplier often lets you down, and the order promised for tomorrow is five months late! The balance between threats and persuasion, correct procedures and a cheerful smile is not easy to achieve. The wheels have to be kept turning, whether it is a simple design problem that has just been outlined or the design of a moon rocket. The temptation of attributing your mistakes and misfortunes to others, however low or high, should be avoided. In each and every department it must be accepted that the other person's knowledge may be more up to date than your own.

DESIGNERS AND SAFETY

Designers should be aware of their contribution to safety, and have full knowledge of the technicalities of safety measures. Machinery guards, dust and fume control and noise control require the same attention at the design office stage as the structure or mechanism from which the hazards originate. All design and drawing office staff should study the Health and Safety at Work Act 1974, and would be wise to obtain the various technical data notes and booklets on safety published by the Department of Employment. These are available from the district offices of Her Majesty's Factory Inspectorate. The various Codes of Practice published by the British Standards Institution should also be available for reference in all design and drawing offices. The essential duty of all designers is to eliminate user-dangers. It is the duty of the designer to be a user. From the safety aspect it is perhaps better if the designer considers that parents and loved ones are also the users. Design without full consideration for safety cannot be termed design at all (see figure 2.15).

At the start of this chapter it was suggested that a designer could be defined as a person who solves problems. He can, however, be

machines, television receivers and egg-timers. Judgement of such claims is left to your own experience or to the valued judgement of previous users.

The purpose of the design task chosen has been to provide a general idea of the people and departments involved in achieving a result. In everyday life there are conflicts of views, opinions and the creator of extra problems or perhaps, in modern terminology, 'negative spin-off'! The type of situation is commonplace, so why not list a few examples similar to the ones shown (see figure 2.16).

Finally it is worth remembering that a good designer, having communicated design thoughts which are put into practice, must accept total responsibility.

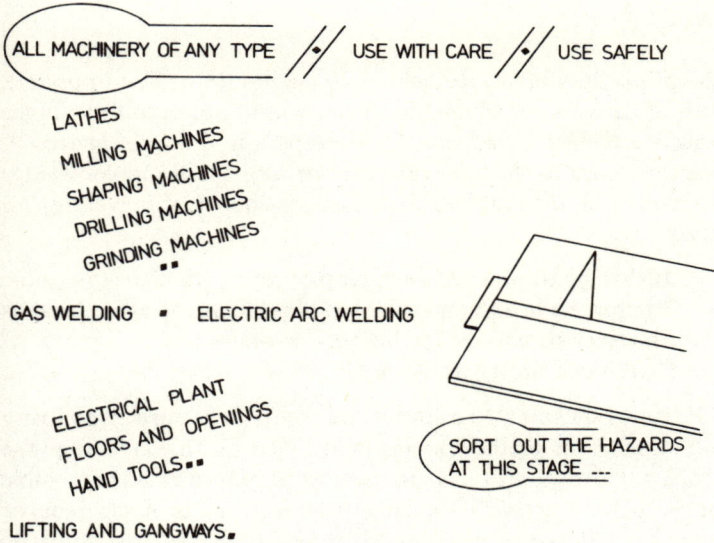

ALL MACHINERY OF ANY TYPE // USE WITH CARE // USE SAFELY

LATHES
MILLING MACHINES
SHAPING MACHINES
DRILLING MACHINES
GRINDING MACHINES ..

GAS WELDING · ELECTRIC ARC WELDING

ELECTRICAL PLANT
FLOORS AND OPENINGS
HAND TOOLS..

LIFTING AND GANGWAYS.

SORT OUT THE HAZARDS AT THIS STAGE ...

SAFE DESIGN
     IS
          GOOD DESIGN !

Figure 2.15

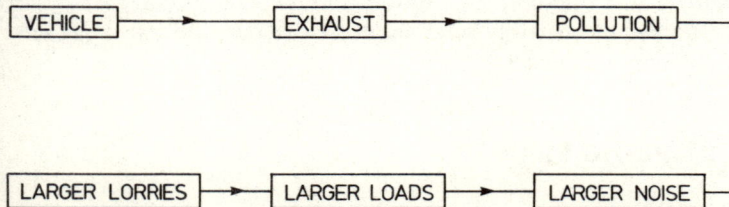

VEHICLE → EXHAUST → POLLUTION

LARGER LORRIES → LARGER LOADS → LARGER NOISE

Figure 2.16

3 Design Elements and Logic

SAFETY

The prime duty of any designer is to make sure, as far as possible, that all dangers are eliminated from within a design task. If the objective danger is reduced, for example the fitting of guards to machinery, and at the same time the operator's skill and knowledge are improved, then at least there is more chance of preventing an injury.

> 'It shall be the duty of every employee . . . to take reasonable care for the health and safety of himself and of other persons who may be affected by his acts or omissions. . . .'
> *Health and Safety at Work Act 1974, Section 7(a)*

Features such as sharp corners on component parts can mostly be 'designed out' on the drawing board. As with all good design the task is not always easy. The designer must reduce as far as possible the need for approach to dangerous features of a machine or assembly. Safety features should be provided that can be easily inspected by persons with authority to do so. Such features should incorporate fail-safe reliable components that are convenient for all those concerned with the machine or assembly and that are difficult to defeat. The student is reminded that the Department of Employment publish a selection of technical notes and booklets dealing with good design practice. Also, reference should be made to existing and forthcoming publications published by the British Standards Institution.

The designer must take all reasonable precautions against failure, and ask himself whether or not his 'reasonable precautions' would hold good in legal proceedings. Another approach is to evaluate the method of failure, should this occur. Will the component or assembly bend, shatter, twist, etc., and also what will be the consequences?

A WORKING PLAN

A really good design can only be achieved by an experienced designer who works under the direction of enlightened management, and by enlightened management we mean people

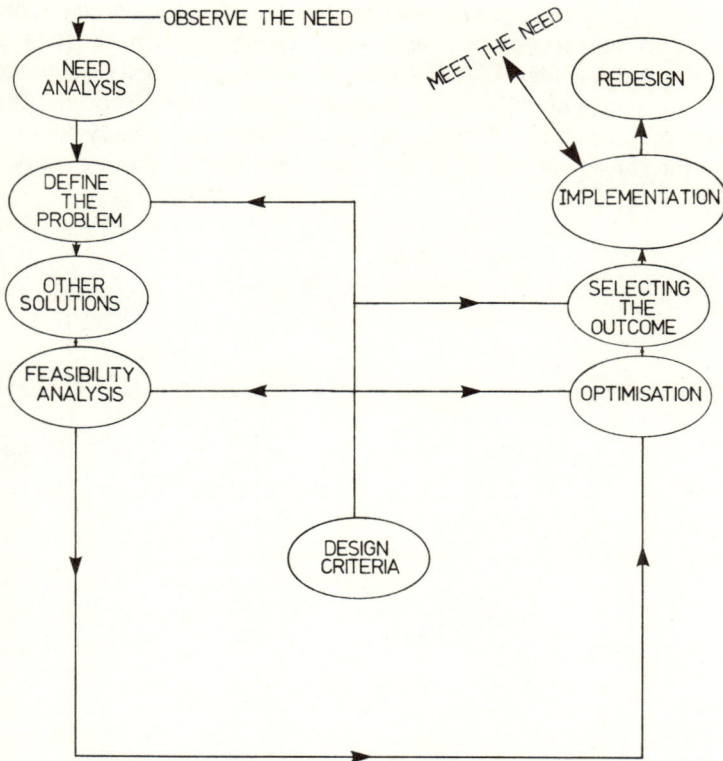

who really care about a product. Very often it is the smaller features of design that are overlooked. A well-designed product is a product that embodies the logical compromise necessary to satisfy the objectives. Efficient performance, ease of use, ease of servicing and value for money are basic requirements of a well-designed product.

'Design methodology' is a term that describes a chosen method of approach to a design problem. The diagram given, together with definitions, is an example of a systematic approach (see figure 3.1). This diagram tends to make the design function look very complicated indeed, and many designers will say that the best approach is to be guided by intuition and apply the pencil to the paper. Although an attempt to apply a systematic approach may not initially provide utopian results, it does at least provide a greater understanding of the complete situation. It helps to clear the fog that can surround the design process.

Here are a few more definitions.

Need analysis, or the analysis of user needs, has the purpose of gathering sufficient information to define the design problem.

Synthesis is the collecting of design thoughts and ideas, and forming them into a systematic arrangement.

Evaluation is the stage of design where the best solution is selected.

Implementation is the final stage of the design process where the pre-production plans are put into action.

Figure 3.1 and its various definitions can also seem rather sophisticated and provide rather an aloof arrangement in the everyday world. However, within the various headings are more commonplace terms and considerations which apply to all designs. To understand and apply the considerations which now follow will enhance your reputation as a designer, for they form the basis of a working plan. Imagine that your design is completed. You are

Figure 3.1		
Need Analysis	Scope and subject background	
Define the Problem	Define the scope, objectives and re-straints relative to needs; show special difficulties to be overcome	
Other Solutions	Provide other solutions, both new and old	
Feasibility Analysis	Analyse reasonable old and new solutions, considering economic, physical and financial aspects	
Implementation	Proceed to produce specifications, models and prototypes for people who need them	
Selecting the Outcome	Judge both the good and bad outcomes of each considered alternative	
Optimisation	View the most likely solutions with respect to the main function	
Design Criteria	Establish sufficient measurable standards for judging the alternative solutions to the design problem (e.g. weight and appearance)	

feeling rather pleased. Now ask yourself a few questions before proceeding to the crucial management meeting.

Consideration (*a*): Does the design meet the user need? Is marketing required?

Consideration (*b*): Check that the parts are assembled easily and if there is need for sub-assemblies.

Consideration (*c*): Is there the correct proportion and symmetry with regard to the function?

Consideration (*d*): Can there be maintenance problems, and are certain components likely to wear? How easy will be the maintenance?

Consideration (*e*): Check the suitability of materials chosen, any alternative materials and the influence of material on the overall appearance. The choice of finish can influence cost—paint, chromium, etc. Colour and texture can enhance a product. (Mention has been made of constraints in the design process, and methods of manufacture used to produce an item are one of the constraints imposed on the designer. It is, therefore, of major importance that the designer should have a broad background knowledge of materials and processes.)

Consideration (*f*): How easy is the product to use or control? How self evident is the use?

Consideration (*g*): Think about the safety aspects outlined earlier in this section.

Consideration (*h*): Cost. Material shortages, price fluctuations and shrinking profits are typical features that affect the design process and are good reasons for the designer to develop a sound approach towards estimating the cost of a product. The accuracy of cost estimates made by the designer in helping him with design decisions need not be totally accurate but they must be accurate enough to provide comparisons. Initially, cost estimating is based upon the cost of materials in a product. Consideration has to be given to the casting and moulding processes, blanking and machining processes, where stock wastage must be expected. There are also the tooling requirements and costs in the case of quantity production.

The overall component manufacturing costs consist of the combined materials, labour and overhead costs.

The selling price of a product can be affected by the novelty of the product or perhaps by competition. However, there must be a financial basis for a fair selling price that will provide a reasonable profit to the manufacturer. With a mass-produced product, the cost of that product decreases as quantity increases, if only because of the cancelling out of tooling costs. But that is another story.

4 The Specification

The engineering model was referred to earlier as a good means of helping to make decisions. The specification can be considered as a means of transmitting decisions. Design information, from the initial concept until final manufacture, should be integrated as much as possible in a logical sequence. Consider the diagram showing a four-stage approach (see figure 4.1). Let us use this approach towards constructing possible specifications for some everyday items.

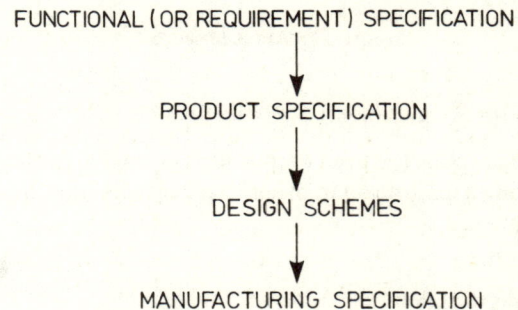

FUNCTIONAL (OR REQUIREMENT) SPECIFICATION

↓

PRODUCT SPECIFICATION

↓

DESIGN SCHEMES

↓

MANUFACTURING SPECIFICATION

Figure 4.1

Item: Safety Razor

Functional (or Requirement) Specification

This would describe the need to remove hair-growth from the human face.

Product Specification

(1) Combined handle and blade-holder;

(2) shielded blade is carried in a disposable dispenser;

(3) blade-holder is offered-up to the dispenser, and a sliding movement into and across the dispenser removes the shielded blade, captivating it in the holder.

Design Scheme

Drawings of the product are completed in order that a critical appraisal regarding competence of the proposed product can be made.

Manufacturing Specification

All instructions for manufacture, assembly, testing and use are prepared. Again, a critical appraisal is essential.

Item: Pocket Flashlight

Functional (or Requirement) Specification

This would describe the need to provide a compact, portable device for producing a light beam in order to give partial illumination in a darkened area.

Product Specification

A moulded plastic container, incorporating a suitable on—off switch, together with a small electric bulb and powered by small batteries.

Try to visualise the time scale involved between initial concept and manufacture, covering the functions of thinking, drawing and making. Again, consider the people involved and not just a single individual. Consider the departments involved, not just the design department. Consider the outside suppliers of materials, tools and equipment, not just the parent manufacturing company. The need to meet customers' requirements and those of the company, customer liaison and good communication with other relevant company departments are covered in detail in later pages.

The following illustrations (see figures 4.2 and 4.3), kindly supplied by MEM Co. Ltd, of Birmingham, show products from their large range of electrical accessories and switchgear: a 1-gang plateswitch, a 2-gang plateswitch, a 13 amp switched socket outlet, and a consumer unit with a range of wiring accessories.

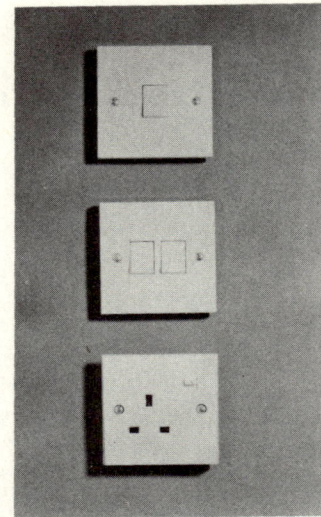

Figure 4.2

Figure 4.3

A catalogue reference to a product may simply state, when referring to a plateswitch: 5A single pole, a.c. only, to BS 3676: 1963. This does not merely mean that they will switch on and off 5 amps 250 volts a.c., but that they have been tested for temperature rises, mechanical life, etc., and have also been subjected to a variety of switching tests on different loads. By this is meant non-inductive and inductive surge loads such as are experienced on fluorescent lighting fittings.

From the foregoing you will gather that small assemblies do not necessarily have small specifications.

A further example of specification information is provided from yet another sector of the design field. The machine illustrated (see figure 4.4) is designed and manufactured by The Bronx Engineering Company Ltd, of Stourbridge, Worcestershire.

Figure 4.4

Bronx precision bar-straightening machines are used in practically every major steelworks throughout the world. They have enabled new standards of straightness and surface finish, together with controlled surface hardness, to be established. The rounding-up effects are most apparent when processing materials of low tensile strength and materials that have good flow characteristics. These bar-straightening machines are particularly well known for their ability to impart a highly polished surface finish to drawn, centreless turned or ground bars. Three features of the machine make this possible: (1) the extremely robust and rigid design of the machine, (2) the specially designed profile of the work rolls, and (3) each work roll is driven from its own separate drive unit. When straightening black 'as rolled' material all primary scale is removed and the bar emerges with a very much improved surface finish.

The prospective customer has a *specific requirement* and therefore the *functional specification* offered by the company is as follows—it describes the *capabilities* or *function* of the machine.

Bright Bar (Centreless Turned or Drawn)

Bars are straightened and polished to a high degree in one pass from a cold drawn or centreless turned process. Bars are precision straightened from end to end.

Black Bars (Hot or 'As Rolled')

Hot-rolled bar ovality rounded up and held to a constant diameter over whole length. Bars precision straightened from end to end. Scale removed and surface finish greatly improved.

Thick-walled Tube

Ideal for precision straightening and rounding up of all types of thick-walled tube.

High Productivity

High throughput speeds of up to 50 m per minute. Automatic handling equipment, allowing one-man operation. Overall efficiency of up to 90 per cent. Very rapid size-changing facilities.

Low Surface Work Hardening

When the machine is fitted with a hydraulic roll load control device, materials which are normally susceptible to work hardening during cold working can be straightened without adversely affecting the physical characteristics.

Having provided a functional specification, the company also provide information regarding the construction and working features of their product, that is, a *product specification*.

Frame

The frame consists of a massively constructed cast iron crown and heavy cast iron base tied together with four steel columns. The crown of the machine supports the top roll and the base of the machine supports the bottom roll.

Main Drive

Each roll is driven from a separate motor. The drive is taken from the motors through vee-belt drives, totally enclosed gear reduction units and mill-type couplings. The two drives are placed at an angle at opposite ends of the machine to approximately conform to the angle of the work rolls.

Rolls

Both rolls are manufactured from a special alloy steel, hardened and ground to a high surface finish on the working areas. The upper roll is provided with a concave profile, the lower roll with a very nearly cylindrical body with tapered ends for easy entry of the bars. The rolls work in heavy duty type, double-taper roller bearings. Special sealing collars are provided to exclude dirt and lubricating solutions.

Roll Adjustment

The bottom roll is fixed in the vertical plane and the top roll is adjustable for bar diameter. This adjustment is by means of a sophisticated fully balanced screw-down worm and wormwheel. Both rolls are provided with angular adjustment, the smaller machines being provided with hand-operated mechanisms. Larger machines are provided with motorised angular adjustment. Where motorised adjustment is provided, adjustment is made from the central operator's control desk.

Guide Bars

A range of lateral guide bars is required to cover bar diameters within the capacity range of the machine. The guide bars are made of a hard-wearing material selected to suit the properties of the bars to be straightened and are easily interchangeable. They are ground on the working face and provided with handwheel adjustment. Individual adjustment is provided at both ends of each guide bar to compensate for any wear. Guide bar holders are adjustable in height. This feature allows the guide bars to be set to the centre-line level of the bars to be straightened. For easy changing of guide bars, the guide bar holders are made to pivot round a column. They can be opened effortlessly and when closed are locked in position.

Lubrication

All points needing lubrication are easily accessible. Lubrication is by hand grease gun. The gears in the reduction gearboxes receive lubrication from the oil bath provided in the entirely closed gearboxes.

Operating Lubrication

A spray pipe fitted near the guide bar directs soluble oil to the workpiece. A motor-driven pump, tank and filtering unit cleanse the solution during operations.

Once again try to visualise the complex involvement of people, materials and manufacturing facilities. Nuts and bolts, motors and vee-belts, letters and telephone calls, are all part of the pattern to provide the right goods at the right price.

When considering the subject of specifications, mention must be made of the British Standards Institution. The institution was founded in 1901 and incorporated by Royal Charter in 1929. The principal objects of the Institution, as set out in the Charter, are to co-ordinate the efforts of producers and users for the improvement, standardisation and simplification of engineering and industrial materials; to simplify production and distribution; to eliminate the waste of time and material involved in the production of an unnecessary variety of patterns and sizes of articles for one and the same purpose; to set up standards of quality and dimensions; and to promote the general adoption of British Standards. The use of structural steel in buildings, unit heads, slotted grub screws, metric keys and keyways, are just a random selection of British Standards which are prefixed with the words 'Specification for . . .'. The British Standards Specification can cover scope, fits, materials, dimensions, forms of construction, working pressures, information to be supplied by the purchasers of an item and many more features. In other words, the British Standard Specification is built around the objectives of the Institution previously mentioned.

THE FEASIBILITY STUDY

The products considered each meet a need and have necessitated the completion of various design tasks. The same can be said for cars, aircraft and oil rigs; in fact anything that meets the user's need. A basic pattern of design criteria is shown in the accompanying diagram (see figure 4.5). You will notice, once again, the all-important mention of costs.

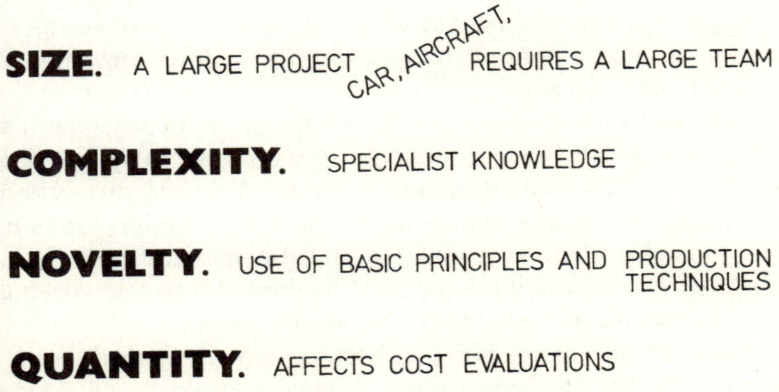

A key part of cost evaluation is the consideration of the quantity of a product to be manufactured. However, the economics of a product are discussed initially when a *feasibility study* is instigated. The aim of the feasibility study is to examine whether or not a design solution will function to the given specification—at an *economic cost*.

Cost comparison with similar products is essential, and this can lead to the feasibility of alternative designs for the products under consideration. The need for alternative designs may be due to the manufacturing facilities available within an organisation. Investment in new manufacturing plant to meet the demands of a new product may be prohibitive for many reasons. Currently, much reference is made to lack of investment in new plant by industry, and the student should reflect upon the effects this may or may not have on new design projects.

SIZE. A LARGE PROJECT *CAR, AIRCRAFT,* REQUIRES A LARGE TEAM

COMPLEXITY. SPECIALIST KNOWLEDGE

NOVELTY. USE OF BASIC PRINCIPLES AND PRODUCTION TECHNIQUES

QUANTITY. AFFECTS COST EVALUATIONS

Figure 4.5

5 Joining of Materials

BOLTS, SCREWS AND NUTS

There is a large variety of standard bolts, nuts and screws which are used to secure various forms of joint. The large number of standard screw threads available emphasises the care that must be taken to ensure that only similar threads are mated. It is essential that the correct spanner or wrench is used to avoid damage to the gripping device, the bolt or nut, and injury to the user. The choice of either screw or rivet in an assembly can depend upon various factors. Typical factors are as follows.

(1) Pressure, humidity and temperature
(2) Expansivity, elasticity and reactivity
(3) Shear, type of load, that is, torsional, impact or vibrational
(4) Type of mating thread, that is, nut or tapped hole, material used, length of thread engagement
(5) Type of assembled or 'joint' material, that is, hardness, elastic characteristics, etc.
(6) Accessibility of joint
(7) Type of assembly, that is, automatic or by hand
(8) Life of assembly
(9) Service requirements, that is, frequent assembly/disassembly

The quantity of fasteners to be used and the final cost of assembly are two further important factors which the designer must always take into consideration.

In more technical terms, the threaded fastener of any type is a device whereby an effective clamping force is produced by applying a torque which is then maintained in service. A varying proportion of the applied torque is used at the nut-face in overcoming friction, in overcoming friction in the thread, and also in extending the bolt. Obviously, the clamping force must be greater than the working load, otherwise the assembled joint would loosen.

The ordinary nut and bolt requires an externally applied torque, but screws with a hexagonal recess, Phillips recess or straight-screw slot require an internally applied torque.

A further consideration is that of the self-tapping screw, which is either thread-forming or thread-cutting. When choosing to use a

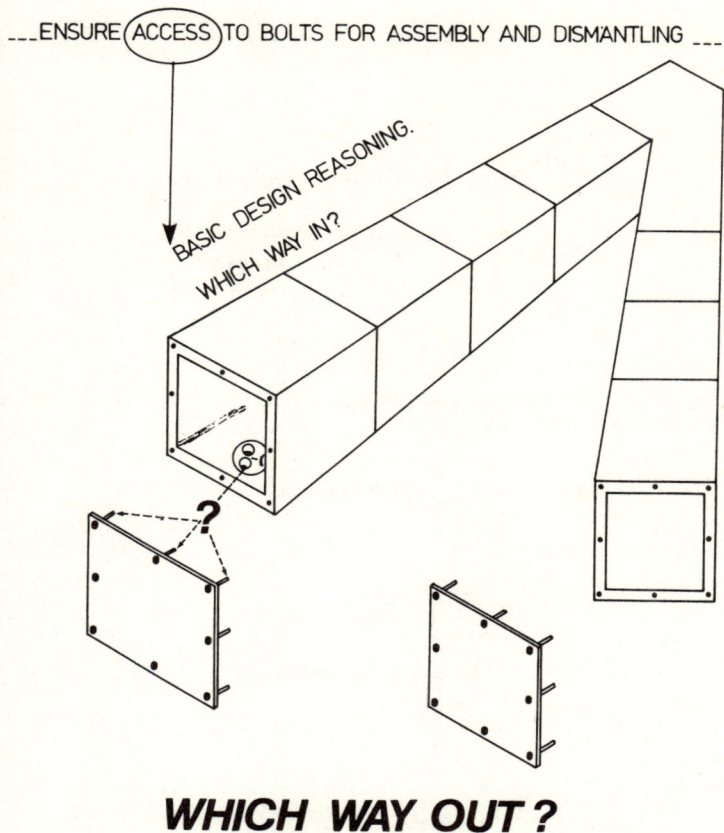

ENSURE ACCESS TO BOLTS FOR ASSEMBLY AND DISMANTLING

BASIC DESIGN REASONING.

WHICH WAY IN?

WHICH WAY OUT ?

Figure 5.1

cut. The black bolt is used in a clearance hole which is usually 1.5 mm larger than the bolt. Fitted bolts, also referred to as precision or turned bolts, have the shank and underhead machined, and the nuts are faced on one side. The holes for such bolts are usually 0.15 mm larger than the bolt, and care should be exercised in the alignment of holes and in the fitting of the bolts.

A suitable washer should be fitted under the nut to avoid damage to the face of the member being tightened. Equally important, the washer increases the frictional grip, thus reducing the tendency to slip and improving the locking action of the nut on the screw thread.

At least two threads on the bolt should protrude through the nut in a bolted assembly. No threads on the bolt should be within the clearance hole, and the contact surfaces of the nut and bolt must be square to the hole. The tapered washer is used to ensure that the contact surface of the bolt-head or nut is square to the hole and is used usually for bolted connections on rolled steel joists, channel, angle and tee-flanges (see figures 5.2 and 5.3).

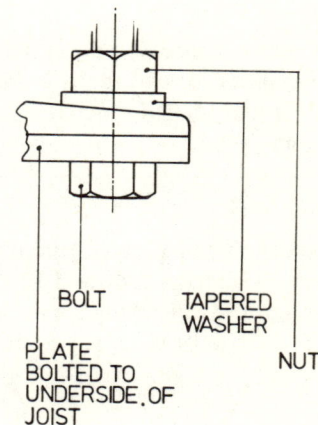

BOLT   TAPERED WASHER   NUT

PLATE BOLTED TO UNDERSIDE OF JOIST

Figure 5.2

self-tapping screw it must be remembered that an excess applied torque will strip the thread from the screwed material. Special proprietary makes of nuts and fastening can be obtained, and can be found incorporated in sheet metal assemblies in many industries as well as in domestic kitchen hardware.

The standard 'hexagon bolt and nut' is to be seen in many bolted assemblies, the hexagonal shape affording a series of positions for tightening. Forged from round steel stock with the head 'upset' by machine, the black bolt (and nut) have only the threads machine

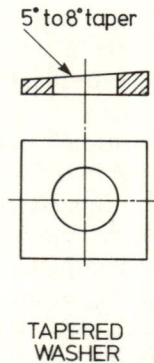

TAPERED WASHER

Figure 5.3

Figure 5.4   ISO metric precision hexagon bolts and screws; these dimensions are laid down in BS 3692: 1967

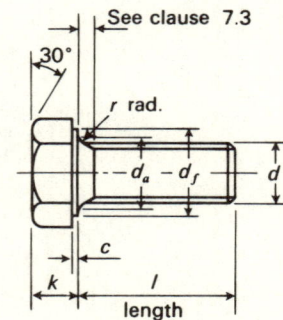

Designation of ISO Metric Threads

Screw threads are designated by the letter M followed by the size of the diameter and the pitch, *both* of which are in mm, for example, $M5 \times 0.8$. For coarse threads, the diameter only is given, e.g. M7; that is, *no pitch* is shown. The complete designation of a thread is shown thus

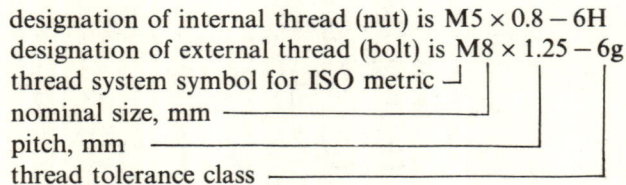

designation of internal thread (nut) is $M5 \times 0.8 - 6H$
designation of external thread (bolt) is $M8 \times 1.25 - 6g$
thread system symbol for ISO metric ⌐
nominal size, mm ———————
pitch, mm ————————
thread tolerance class —————

A fit between a pair of threaded parts is indicated by the internal (nut) tolerance class designation followed by the external thread (bolt) class designation, the two being separated by a stroke, for example, $M8 \times 1.25 - 6H/g$. (See figures 5.4 to 5.6 and table 5.1.)

Length of Thread

The length of thread is given by the following formulae. Note that the length of thread is greater for all bolts when compared to the inch sizes of BS Whitworth and BSF and greater in longer lengths when compared to UNC and UNF.

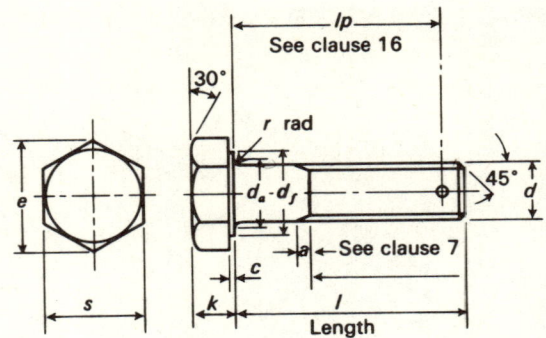

Figure 5.5   Hexagon head screw, washer faced

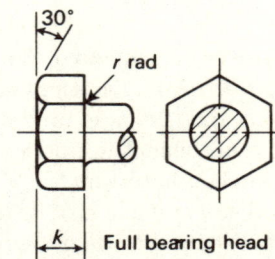

Figure 5.6   Alternative type of head permissible on bolts and screws

Table 5.1

Nominal Size, d	Pitch of Thread	Width across Flats		Thickness of Nut		Height of Head of Bolt		Width across Corners	
		max	min	max	min	max	min	max	min
M4	0.7	7.00	6.85	3.20	2.90	2.925	2.675	8.10	7.74
M5	0.8	8.00	7.85	4.00	3.70	3.650	3.35	9.20	8.87
M6	1	10.00	9.78	5.00	4.70	4.15	3.85	11.50	11.05
M8	1.25	13.00	12.73	6.50	6.14	5.65	5.35	15.00	14.38
M10	1.5	17.00	16.73	8.00	7.64	7.18	6.82	19.60	18.90
M12	1.75	19.00	18.67	10.00	9.64	8.18	7.82	21.90	21.10
M14	2	22.00	21.67	11.00	10.57	9.18	8.82	25.40	24.49
M16	2	24.00	23.67	13.00	12.57	10.18	9.82	27.70	26.75
M18	2.5	27.00	26.67	15.00	14.57	12.215	11.785	31.20	30.14
M20	2.5	30.00	29.67	16.00	15.57	13.215	12.785	34.60	33.53
M22	2.5	32.00	31.61	18.00	17.57	14.215	13.785	36.90	35.72
M24	3	36.00	35.38	19.00	18.48	15.215	14.785	41.60	39.98
M27	3	41.00	40.38	22.00	21.48	17.215	16.785	47.3	45.63
M30	3.5	46.00	45.38	24.00	23.48	19.26	18.74	53.1	51.28

Proportions which can be used when drawing: across the flats 1.5d; thickness of nut 0.8d; height of bolt head 0.7d.

Nominal Length of Bolt	Length of Thread
Up to and including 125 mm	2 × dia. + 6 mm
Over 125 mm up to and including 200 mm	2 × dia. + 12 mm
Over 200 mm	2 × dia. + 25 mm

Note that bolts that are too short for minimum thread lengths are threaded and designated as screws (see table 5.2).

Head Marking

The marking in figure 5.7a indicates that the bolt is made of ISO grade 8.8 steel. The marking in figure 5.7b is in the form of a code system based on a clockface, with a single dot indicating 12 o'clock. The second mark—a bar—indicates the grade; for example, in the case of a grade 8 nut the bar is at 8 o'clock on top of the nut. Such marks are indented.

Ends of Bolts and Screws

The ends of bolts and screws may, at the discretion of the manufacturer, be finished with either a 45° chamfer to a depth of slightly more than the depth of thread, or by a radius approximately equal to 1.25 times the nominal diameter of the shank. When bolts and screws are made with rolled threads, the lead formed at the end of the bolt by the thread-rolling operation may be regarded as providing the chamfer to the end, and no other machining operation is required, provided that the end is reasonably square with the centre-line of the shank. Alternative types of permissible end are shown in figure 5.8.

Table 5.2 Shortest Preferred Lengths (mm) Designated as Bolts

Diameter		
	M5	20
	M6	25
	M8	30
	M10	35
	M12	40
	M14	45
	M16	45
	M18	50
	M20	55
	M22	60
	M24	65
	M27	70
	M30	80
	M33	85
	M36	90
	M39	100
	M42	110

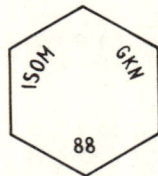

(a) Head marking

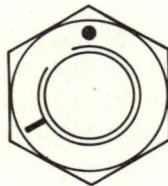

(b)

Figure 5.7

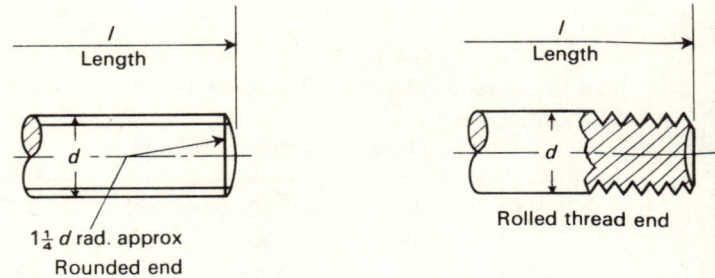

Figure 5.8

Nuts

Nuts shall be countersunk on the bearing face at an included angle of 120° ± 10°. The diameter of the countersink must not exceed the nominal major diameter of the thread (see figure 5.9).

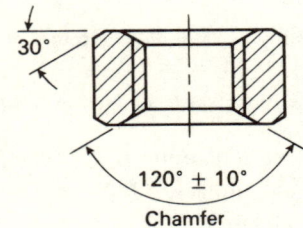

Figure 5.9 Enlarged view of countersink

ISO Metric Precision Hexagon Slotted Nuts and Castle Nuts

See figures 5.10 to 5.12 and table 5.3.

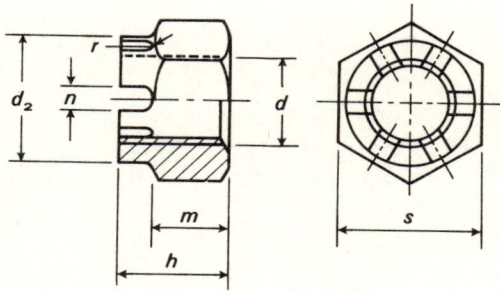

Figure 5.10   Castle nut, sizes M12 to M39 only (six slots)

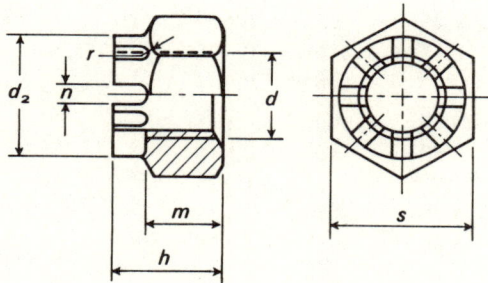

Figure 5.11   Castle nut, sizes M42 to M68 only (eight slots)

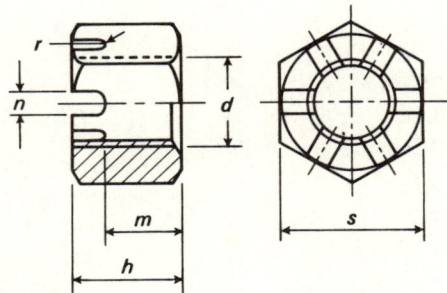

Figure 5.12   Slotted nut, sizes M4 to M39 only (six slots)

Machine Screws

BS 4183: 1970 gives details of the following metric machine screws.

> pan head
> countersunk head
> raised countersunk head
> cheese head
> slotted and Posidriv recessed head machine screws
> slotted machine screws only

All head dimensions are now directly related to the basic screw diameter.

Screws are threaded to permit a screw ring gauge to be screwed by hand to within a distance from the underside of the head not exceeding $2.5 \times$ the pitch for diameters up to and including 25 mm, and $3.5 \times$ the pitch for diameters over 25 mm.

Metric machine screws are designated as follows: material and style of screw or nut followed by the diameter $\times$ pitch $\times$ length and BS number; for example, steel cheese head screw: $M6 \times 1.0 \times 20$ mm to BS 4183. See figure 5.13 and table 5.4.

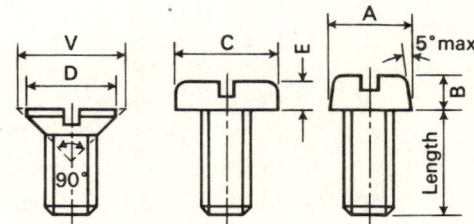

Figure 5.13

Dimensions of ISO Metric Hexagon Socket Screws

To BS 4168: 1967; threads to BS 3643 Class 6g; all dimensions in mm (see figures 5.14 and 5.15 and tables 5.5 to 5.7).

Table 5.3

1	2		3		4		5		6
Nominal Size and Thread Diameter d	Diameter d_2		Thickness h		Lower Face of Nut to Bottom of Slot m		Width of Slot n		Radius $(0.25n)$ r
	max	min	max	min	max	min	max	min	min
M4	—	—	5	4.70	3.2	2.90	1.45	1.2	0.3
M5	—	—	6	5.70	4.0	3.70	1.65	1.4	0.35
M6	—	—	7.5	7.14	5	4.70	2.25	2	0.5
M8	—	—	9.5	9.14	6.5	6.14	2.75	2.5	0.625
M10	—	—	12	11.57	8	7.64	3.05	2.8	0.70
M12	17	16.57	15	14.57	10	9.64	3.80	3.5	0.875
(M14)	19	18.48	16	15.57	11	10.57	3.80	3.5	0.875
M16	22	21.48	19	18.48	13	12.57	4.80	4.5	1.125
(M18)	25	24.48	21	20.48	15	14.57	4.80	4.5	1.125
M20	28	27.48	22	21.48	16	15.57	4.80	4.5	1.125
(M22)	30	29.48	26	25.48	18	17.57	5.80	5.5	1.375
M24	34	33.38	27	26.48	19	18.48	5.80	5.5	1.375
(M27)	38	37.38	30	29.48	22	21.48	5.80	5.5	1.375
M30	42	41.38	33	32.38	24	23.48	7.36	7	1.75
(M33)	46	45.38	35	34.38	26	25.48	7.36	7	1.75
M36	50	49.38	38	37.38	29	28.48	7.36	7	1.75
(M39)	55	54.26	40	39.38	31	30.38	7.36	7	1.75
M42	58	57.26	46	45.38	34	33.38	9.36	9	2.25
(M45)	62	61.26	48	47.38	36	35.38	9.36	9	2.25
M48	65	64.26	50	49.38	38	37.38	9.36	9	2.25
(M52)	70	69.26	54	53.26	42	41.38	9.36	9	2.25
M56	75	74.26	57	56.26	45	44.38	9.36	9	2.25
(M60)	80	79.26	63	62.26	48	47.38	11.43	11	2.75
M64	85	84.13	66	65.26	51	50.26	11.43	11	2.75
(M68)	90	89.13	69	68.26	54	53.26	11.43	11	2.75

Note Non-preferred sizes shown in parentheses. For widths across flats and corners, see table 5.1.

Table 5.4

Nominal size	Pitch	Countersunk Head			Pan Head				Cheese Head			
		dia of head			dia. of head C		depth E		dia. of head A		depth B	
		V max (sharp)	D min	max land on min dia	max	min	max	min	max	min	max	min
M3	0.50	6.00	5.25	0.37	6.00	5.70	1.80	1.66	5.50	5.20	2.00	1.86
M4	0.70	8.00	7.00	0.50	8.00	7.64	2.40	2.26	7.00	6.64	2.60	2.46
M5	0.80	10.00	8.75	0.62	10.00	9.64	3.00	2.86	8.50	8.14	3.30	3.12
M6	1.00	12.00	10.50	0.75	12.00	11.57	3.60	3.42	10.00	9.64	3.90	3.72
M8	1.25	16.00	14.00	1.00	16.00	15.57	4.80	4.62	13.00	12.57	5.00	4.82

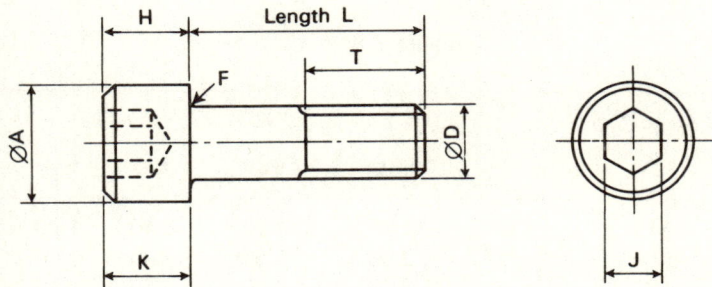

Figure 5.14

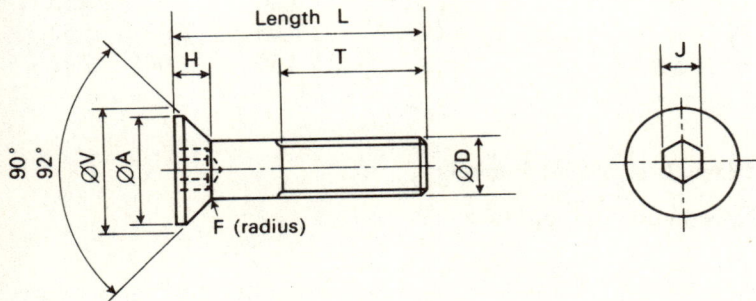

Figure 5.15

Table 5.5

Thread Size D	A max	H max	J nom	K min	F max
M3	5.5	3	2.5	1.3	0.30
M4	7.0	4	3.0	2.0	0.35
M5	8.5	5	4.0	2.7	0.35
M6	10.0	6	5.0	3.3	0.40
M8	13.0	8	6.0	4.3	0.60
M10	16.0	10	8.0	5.5	0.60
M12	18.0	12	10.0	6.6	1.10

Table 5.6   Thread Length, T (cap and countersunk heads)

Nominal Length, L		Formula
over	up to	
—	125	2D + 6
125	200	2D + 12
200	—	2D + 25

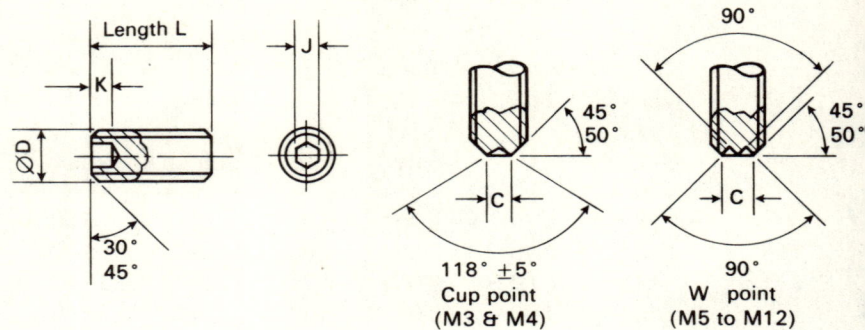

Figure 5.16

Table 5.7

Thread Size	V*	A	H	F	J
D	max	min	max	max	nom
M3	6.72	5.82	1.86	0.4	2.0
M4	8.96	7.78	2.48	0.4	2.5
M5	11.20	9.78	3.10	0.4	3.0
M6	13.44	11.73	3.72	0.6	4.0
M8	17.92	15.73	4.96	0.7	5.0
M10	22.40	19.67	6.20	0.8	6.0
M12	26.88	23.67	7.44	1.1	8.0

Table 5.8

Thread Size	J	K	C	
D	nom	min	max	min
M3	1.50	1.20	1.40	1.00
M4	2.00	1.50	2.00	1.60
M5	2.50	2.00	2.50	2.10
M6	3.00	2.40	3.00	2.60
M8	4.00	3.20	5.00	4.52
M10	5.00	4.00	6.00	5.52
M12	6.00	4.80	8.00	7.42

* Dimension V is the theoretical diameter of head to sharp corners and is given for design purposes only

Details of Set Screws

See figure 5.16 and table 5.8.

Details of Metric Washers

See figure 5.17 and table 5.9.

I.S.O washers

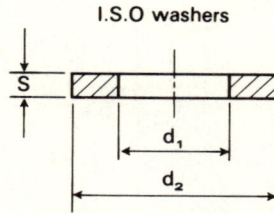

Figure 5.17   Details of metric washers

Table 5.9

Thread Diameter	Clearance Hole		Diameter of Washer		Thickness
(mm)	d_1		d_2		S (mm)
	bright	black	normal	large	
3	3.2	—	7.0	—	0.5
4	4.3	—	9.0	—	0.8
5	5.3	5.5	10.0	—	1.0
6	6.4	6.6	12.5	—	1.5
8	8.4	9.0	17.0	21.0	1.5
10	10.5	11.0	21.0	24.0	2.0
12	13.0	14.0	24.0	28.0	2.5
16	17.0	18.0	30.0	34.0	3.0
20	21.0	22.0	37.0	39.0	3.0
24	25.0	26.0	44.0	50.0	4.0
30	31.0	33.0	56.0	60.0	4.0
36	37.0	39.0	66.0	72.0	5.0

Marking

Bolts and screws of diameter 6 mm and larger shall be identified as being metric by having either the symbol ISOM or M embossed or indented on the top of the head.

Bolts and screws turned from hexagon bar may have the ISO metric symbol M indented or rolled into one of the hexagon flats (see figure 5.18).

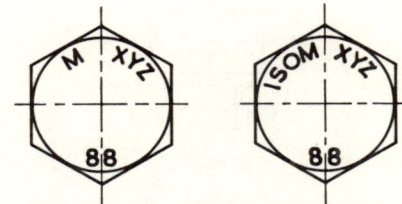

Examples of marking of forged products
M or ISOM = ISO metric identification
XYZ = manufacturer's identification
(trade) marking

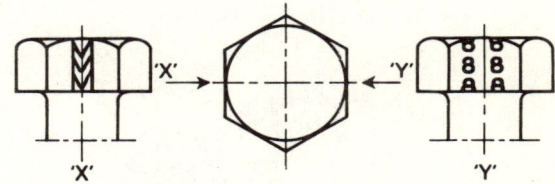

Example of marking of bar turned product

Figure 5.18

Basic Locking Devices

See figures 5.19 to 5.23.

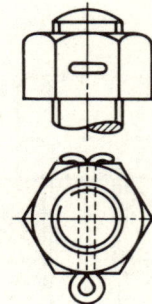

Figure 5.19   Slit pin through nut and bolt

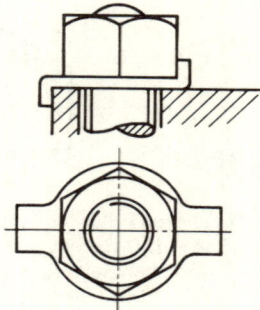

Figure 5.20   Tab washer

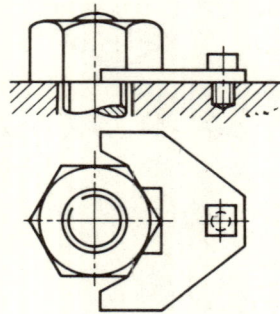

Figure 5.21   Locking plate

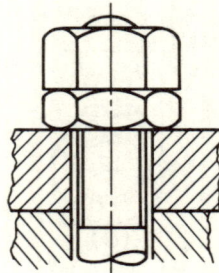

Figure 5.22   Use of locknut

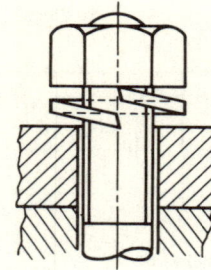

Figure 5.23   Spring washer

RIVETED JOINTS

The strength of fasteners and connected members, when well designed, should be in balance. Members and fasteners can be subjected to stresses in tension, shear, torsion, compression or bending, and a combination of these stresses (see figure 5.24).

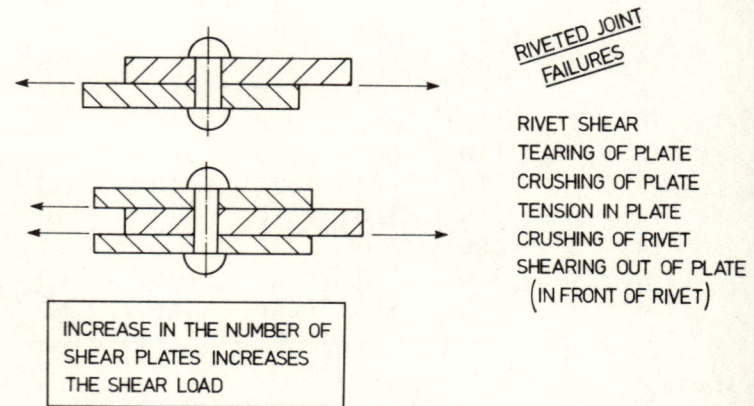

RIVETED JOINT FAILURES

RIVET SHEAR
TEARING OF PLATE
CRUSHING OF PLATE
TENSION IN PLATE
CRUSHING OF RIVET
SHEARING OUT OF PLATE
(IN FRONT OF RIVET)

INCREASE IN THE NUMBER OF SHEAR PLATES INCREASES THE SHEAR LOAD

Figure 5.24

There are instances where the fastener design is such that it will fail under certain conditions as a safeguard for other parts of an assembly. Rivets are used when it is required to connect two or more sections of materials together permanently. They are used in their many forms and sizes in a large range of industries.

The materials for rivet manufacture include mild steel, aluminium, duralumin, copper and brass. A selection of types of rivet head are shown (see figure 5.25). For more detailed information, reference should be made to the relevant British Standard.

Where two or more plates are joined together by riveting, the connection is termed a riveted joint. With the *lap joint* the plates

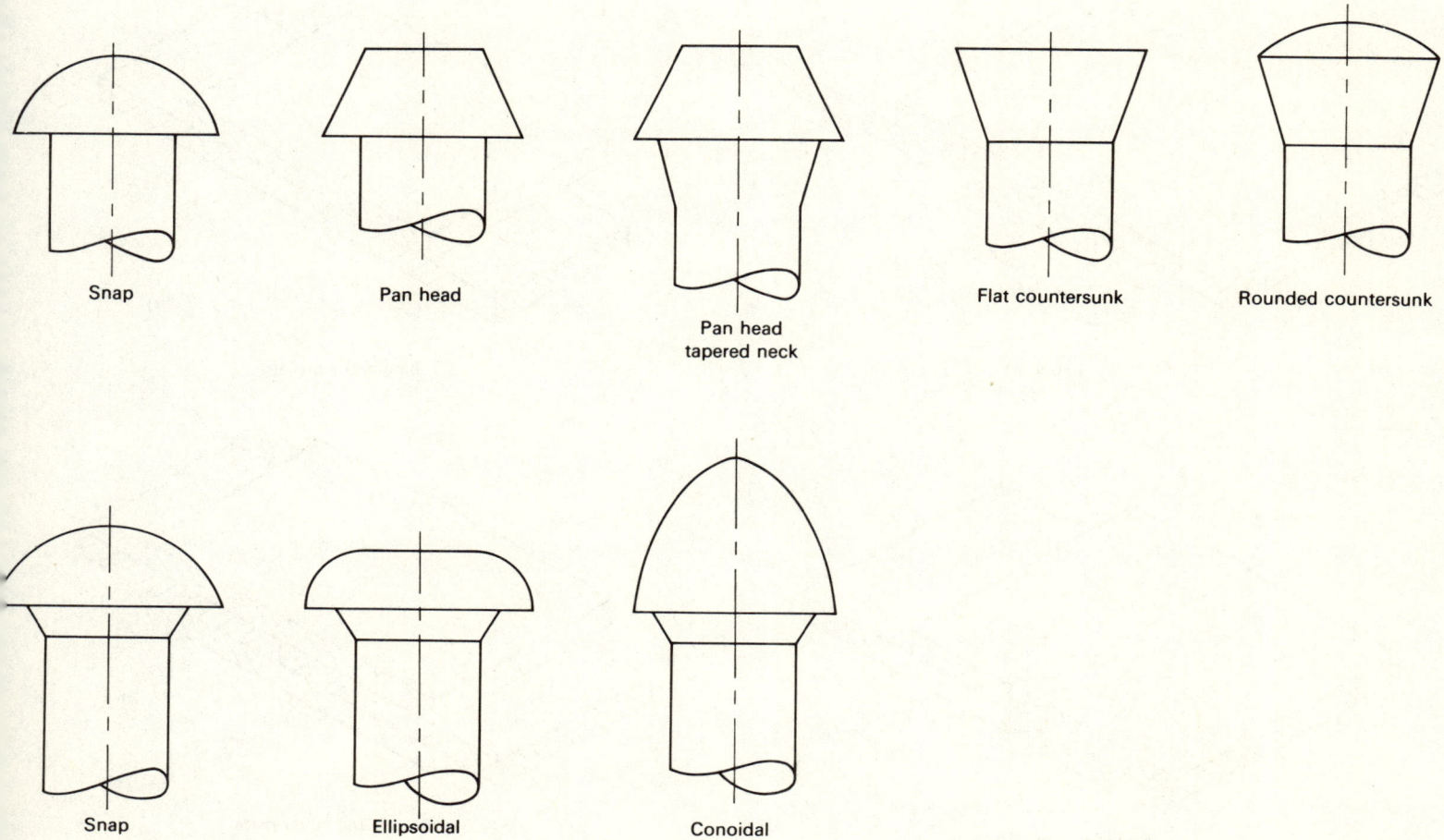

Snap

Pan head

Pan head
tapered neck

Flat countersunk

Rounded countersunk

Snap

Ellipsoidal

Conoidal

Figure 5.25   Types of rivet head

overlap each other by a distance termed an 'overlap' (see figures 5.26 and 5.27).

With the *butt joint*, the main plates to be joined together are butted against each other and the line of junction is then covered by one or more cover straps or plates. The diagrams show a butt joint with one cover plate, and a butt joint with two cover plates (see figure 5.28).

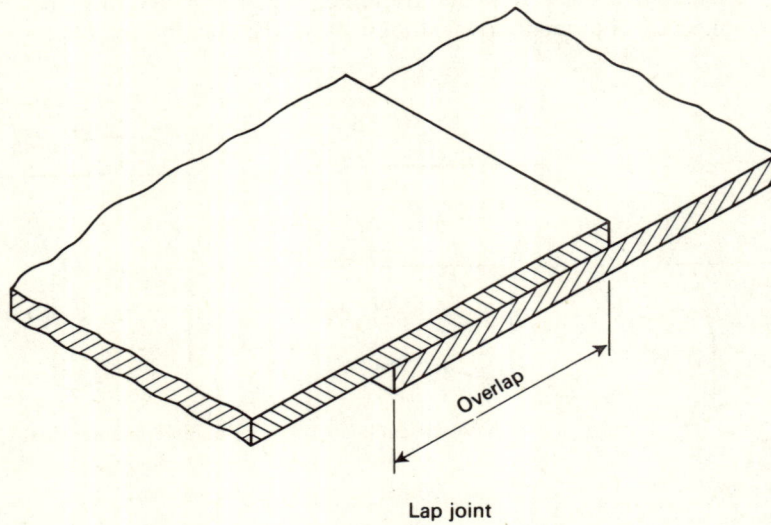

Lap joint

Figure 5.26

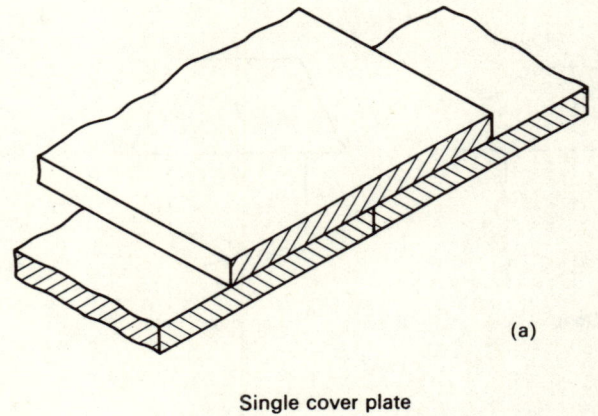

(a)

Single cover plate

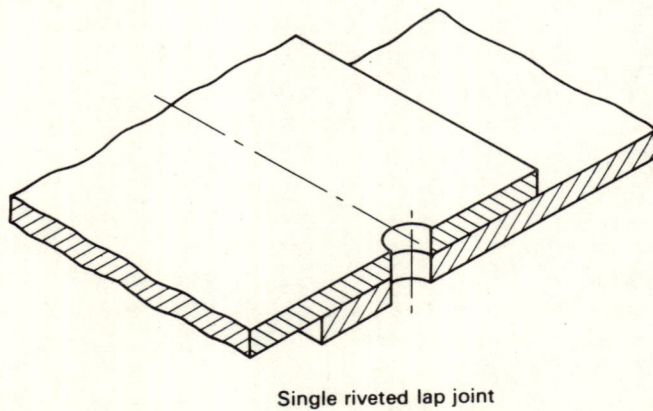

Single riveted lap joint

Figure 5.27

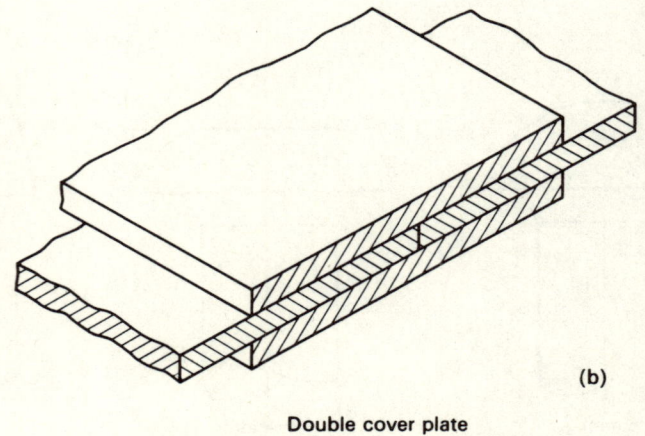

(b)

Double cover plate

Figure 5.28

Rivets can be arranged in several different ways, depending upon the stresses taken by the joint under working conditions. Single, double and treble riveting are the most common patterns. The pattern or type of riveting refers to the number of rows of rivets counted from the joint of one plate. In a single lap joint there is one row. In a single-riveted butt joint there are two rows, one in each plate. A double-riveted butt joint has four rows, two in each plate. A treble-riveted butt joint has six rows, three in each plate. See figure 5.29.

The next diagram (see figure 5.30) shows how rivets are used in structural assemblies, and it will be noted how use is made of rolled steel joists and channel and angle sections.

The riveting operation can be performed hot or cold. With the use of new riveting tools the cold method is often used, offering speed and efficiency. The length of rivet required is different for the hot and cold methods.

A common method of forming the head on a rivet is by peening. The rivet material is swaged outwards and downwards into contact with the sides of the hole in which the rivet is to be assembled. The remainder of the material is swaged into the head form.

A selection of dimensional terms applied to riveting is as follows—as already mentioned, either lap or butt joints are generally used in riveted joint design. The *gauge line* is the centreline through a row of rivets drawn parallel to the joint edge of the plates. The distance from the edge of the plate to the first gauge line is termed the *edge distance*. *Pitch* is the distance between rivets. *Transverse pitch* is the distance between gauge lines.

In order to obtain a correct balance of strength, the material for the rivet should be of about the same strength as the material to be riveted. The correct choice of rivet pitch will help to avoid the tearing of the plate between the rivets. As a general guide, the minimum spacing of rivets is three times the rivet diameter. Enough space must be left between rivets to allow head-forming operations. However, if the pitch is too large, buckling of the metal between the rivets may occur. In order to prevent the shearing, buckling or tearing of the plate between the rivet hole and the edge of the plate, the edge distance should be a minimum of $1\frac{1}{2}$ times the diameter of the hole.

The hole size for a rivet should be as small as possible and yet

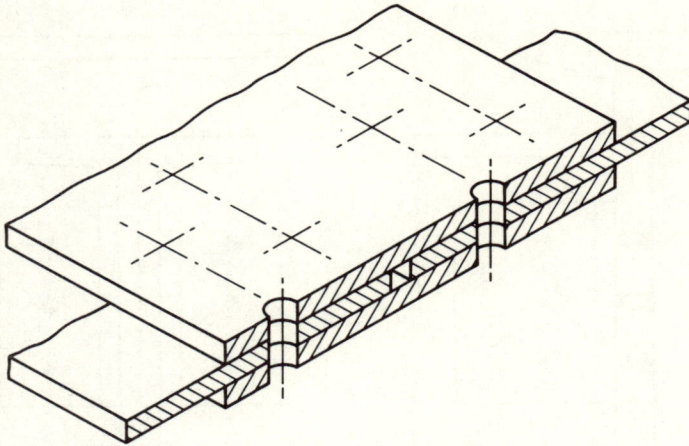

Double riveted butt joint

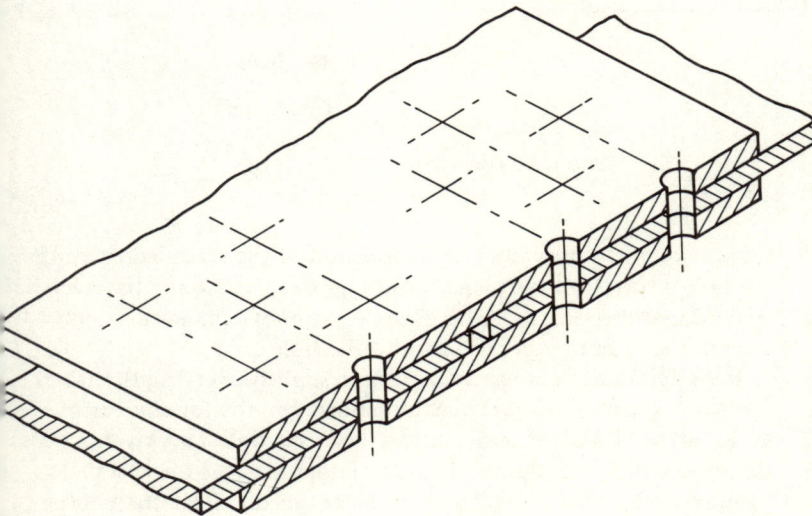

Treble riveted butt joint

Angle

Gusset plate

⊕ Rivet

Flat plate

Gusset connection

Beam end connection

● Bolts

⊕ Rivets

Built up box girder

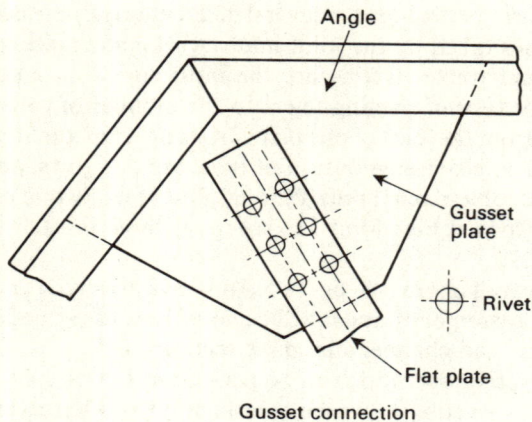

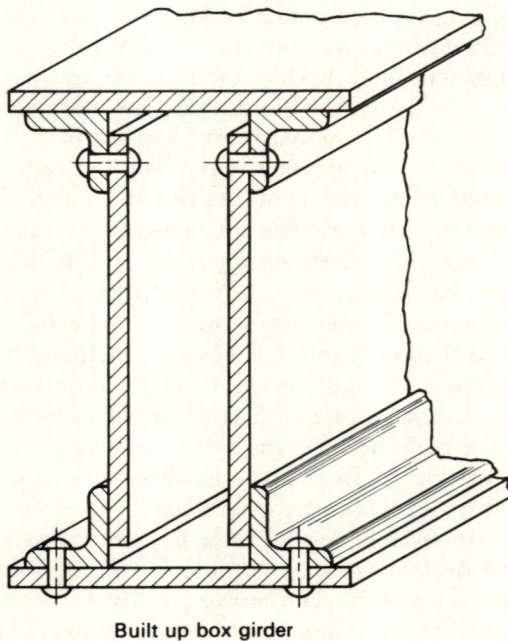

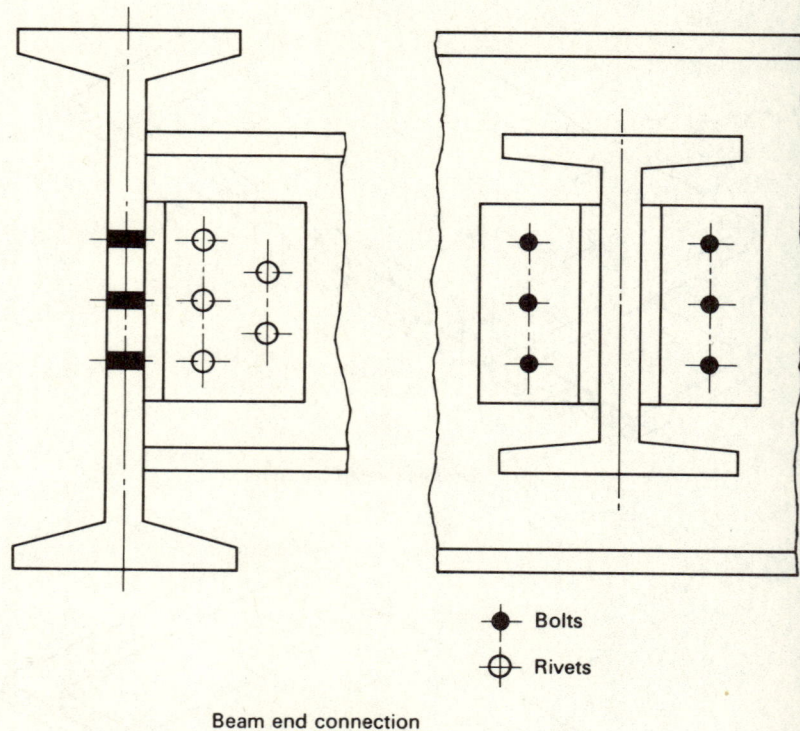

large enough to provide speedy insertion of the specified rivet. If the hole is too large, the plates may bulge or separate and the rivets may bend or become loose because they cannot expand sufficiently during the riveting operation to fill the hole.

It is advisable to choose a rivet length slightly over-length rather than too short. There are recommended lengths for the various types of head and also clearances between hole and rivet. The diameter of the rivet should always be larger than the thickest plate. However, the rivet should not be larger in diameter than three times the thickness of the thinnest plate. It must be remembered that a riveted joint cannot be as strong as a solid sheet or plate.

Figure 5.30

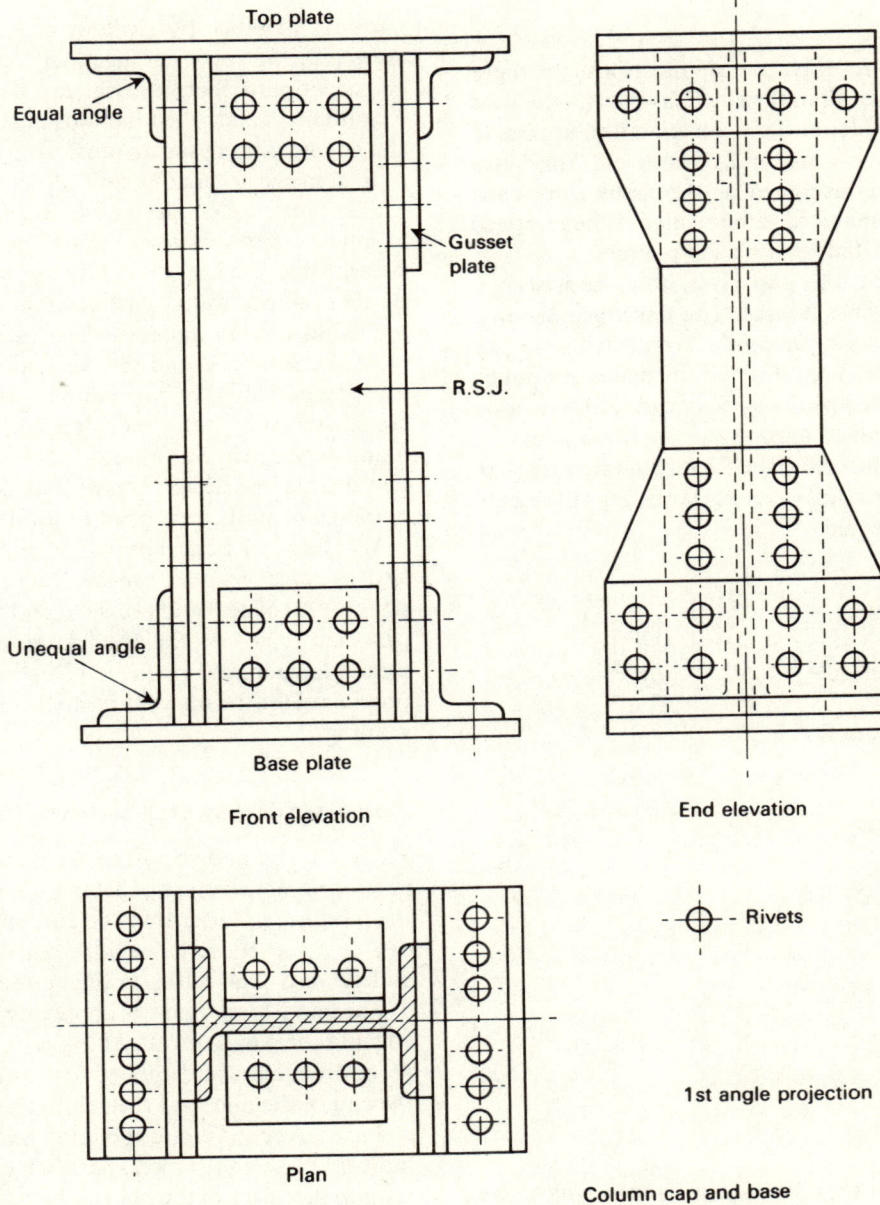

Top plate

Equal angle

Gusset plate

R.S.J.

Unequal angle

Base plate

Front elevation

End elevation

Plan

Rivets

1st angle projection

Column cap and base

Figure 5.30 (*continued*)

'Light Type' Riveting

As a fixing method for certain industrial applications, the 'light type' of riveting can be cheaper than threaded fasteners, when the process of automation is considered. This consideration applies to the tubular or semi-tubular rivets, and bifurcated rivets. This latter type are solid rivets which are machined on the shank to provide two prongs which pierce the material to be riveted. When applied to a material such as leather, they can be self-piercing.

Another type of 'light' rivet is the 'pop' rivet, which consists of a hollow rivet mounted on to a headed pin. The pin is gripped in a special tool and the rivet is positioned in a pre-drilled hole (see figure 5.31). As the tool and pin are retracted the plates are pulled tightly together, and the headed pin swages a head in the rivet on the blind side. At a predetermined tension the pin breaks, leaving only the pin head in the fully formed rivet. (With a number of such tools used in conjunction with an indexing table, an automated riveting process could be effected.)

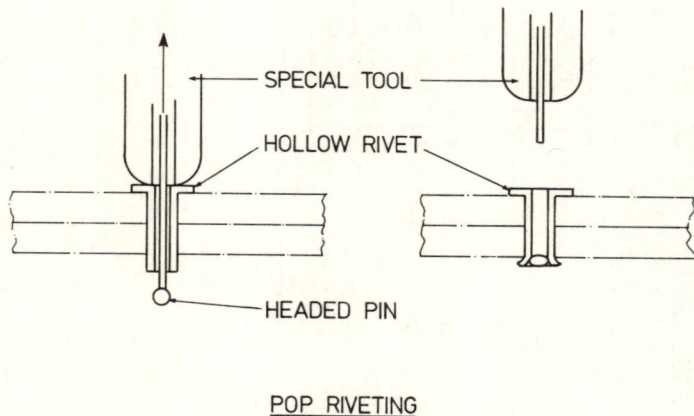

POP RIVETING

Figure 5.31

WELDED JOINTS

A fabricated component consists of several items which are fastened together by riveting or bolting, by the application of adhesives or by welding. An important feature of welded construction is that material can be positioned where it is really required to satisfy the functions of strength and performance, and the engineering designer must always be aware of this.

In contrast to brazing and soldering, where the materials to be joined never reach the liquid stage, welding is achieved by the joining of parts in their liquid form at the joint. Some of the present-day welding processes are highly sophisticated and have been developed to suit particular materials or certain applications.

Fusion welding processes are mainly used in the fabrication of welded assemblies, and the designer and draughtsman must have knowledge of the procedures for these various processes. Consideration of the most economical means of manufacture, load-carrying capacities and function of the finished fabrication, will dictate the design layout and detailed design of the various component parts and types of joint.

Welding is a large, specialised subject, but can be divided into two groups: *pressure welding*, where parts are joined by the application of heat and pressure; and *fusion welding*, where separate parts are joined by the application of heat and a filler rod at the joint. Consideration will be given in particular to the two most popular of the fusion welding methods, namely gas welding and arc welding.

Fusion Welding by Oxy-acetylene Heat Source

When oxygen and acetylene are mixed together and burned they produce a temperature of 3500°C. Special torches are used to burn the mixture, in order that the correct ratio between the two gases can be achieved. A flame with equal volumes of gas is the most widely used. The set-up is illustrated (see figure 5.32).

The hand-torch flame is applied to the joint in order to melt the edges of the parent metal. At the same time a metal filler rod is held in the other hand and applied to the welding zone. The result is that the end of the filler rod melts into the molten pool. The flame and rod are advanced along the joint, and the required size of weld is built-up between the metal parts which are to be joined. Butt joints require the edges of the plate to be prepared before welding and, in

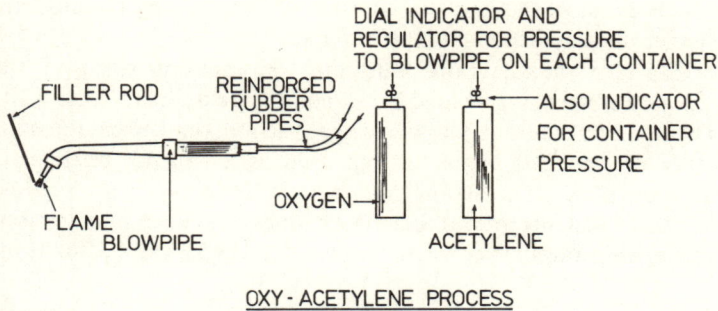

FILLER ROD
REINFORCED RUBBER PIPES
DIAL INDICATOR AND REGULATOR FOR PRESSURE TO BLOWPIPE ON EACH CONTAINER
ALSO INDICATOR FOR CONTAINER PRESSURE
FLAME
BLOWPIPE
OXYGEN
ACETYLENE

OXY-ACETYLENE PROCESS

Figure 5.32

INPUT
TRANSFORMER
OUTPUT
ELECTRODE
COMPONENTS
METAL TABLE

ELECTRICAL ARC WELDING

Figure 5.33

every case, all edges should be clean. There are various types of filler rods available. For example, mild steel rods are used when welding mild steel fabrications, whereas cast iron is best welded using a bronze filler rod and a special flux.

Fusion Welding by Metal Arc (Electric Arc Welding)

With this method an electrical circuit is completed when the electrode makes close contact with the work. A low-voltage high-current arc, struck between a metal electrode (coated with flux) and the materials to be joined, produces a heating at each end of the arc. The end of the welding rod is thus melted together with the metals to be joined. The temperature is greater than that produced by the oxy-acetylene flame, and protection from oxidation is afforded by the gas shield produced by the flux coating on the electrode. The flux forms a molten surface over the weldpool, and becomes brittle enough to chip away after solidifying. Different electrode diameters and different currents are used to suit the differing metal thicknesses. This method of fusion welding is widely used in industry, because it has the advantages of low capital and running costs. The set-up is illustrated (see figure 5.33).

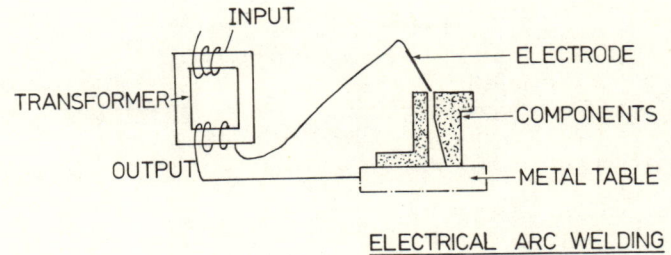

Fusion Welding with Gas Shield

(1) Tungsten Inert Gas (TIG), Argon Arc

With this method an arc is struck between a non-consumable tungsten electrode, and an external metal rod filler is used. The protection from contamination, that is, from reaction with atmospheric oxygen, is provided by a flow of argon. Welds of high quality are obtained.

(2) Inert Gas-shielded Metal Arc (MIG), Sigma

With this method a consumable metal electrode is used, without the use of an external filler rod. With the welding of mild steel by this method, contamination protection can be provided by using carbon dioxide gas, thus reducing costs.

Resistance Welding

Metals vary in their ability to carry an electric current; steel, for example, offers a considerable resistance to the passage of an electric current. The production of domestic appliances and motor car bodies requires a fast, highly productive low-cost welding method, and resistance welding is widely used. The diagram (see figure 5.34) shows the application of resistance welding.

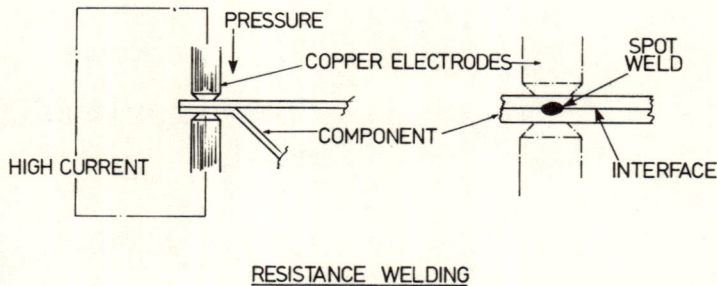

RESISTANCE WELDING

Figure 5.34

The high current is offered small resistance by the copper electrodes, but meets a large resistance at the interface of the joint. The result is the rapid heating of the joint at the interface of the two metal sheets, with fusion taking place under pressure. A spot weld is the result of the process and can be likened to a rivet. Because air is squeezed out of the contact area, no flux is necessary.

Welding techniques have made great advances during recent years, aiding the mass production requirements of various industries. Production of welded joints by solid-phase welding, using friction/pressure, explosive impact and ultrasonic methods, is now occurring in industry. The success of the components produced in the welding shop still depends to a large extent on the careful preparation and execution of the necessary drawings, with the relevant welding symbols.

Fabrication Design Considerations

The engineering designer who decides to fabricate by welding will have to consider factors other than the basic joining of components. There are, for example, the location of components in relation to each other, and the transmission of forces due to the working conditions of the assembly. In terms of load-carrying capacity, the general layout and the shape and size of welds, together with the properties of parent and weld metal, will all influence the design. From this it follows that the method of producing the welds, the quality of workmanship together with

possible distortions and/or internal stresses during cooling, are further considerations for the designer.

Consider the diagrams outlining the options open to the designer when using welded construction (see figures 5.35 to 5.38). The first two diagrams show a flanged section, one of bent form the other of standard angle section, used as a stiffener on a plate component. Such an arrangement would be necessary for a bolted or riveted stiffener and, at first glance, one could keep such sections when using a welded construction (see figures 5.35 and 5.36). If the

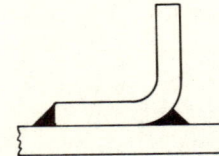

Figure 5.35

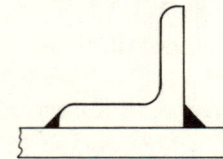

Figure 5.36

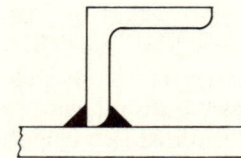

Figure 5.37

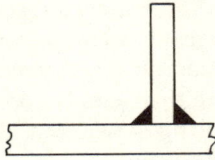

Figure 5.38

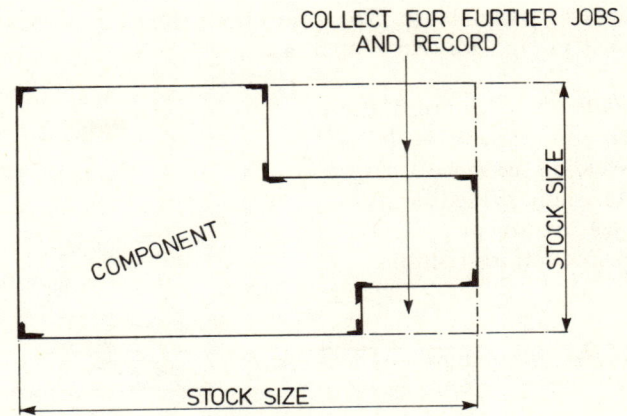

Figure 5.39

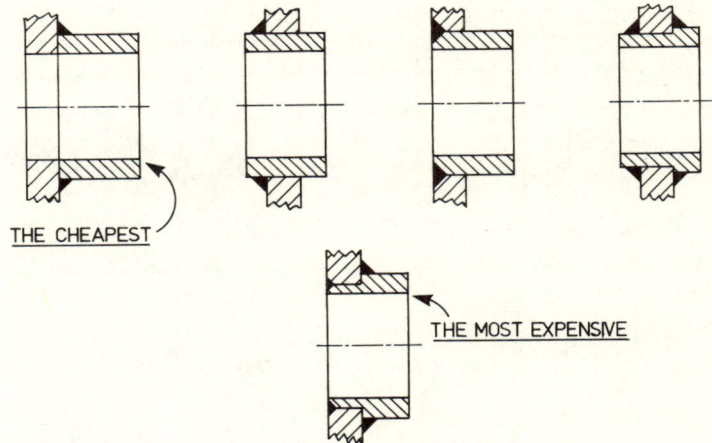

Figure 5.40

maximum stiffness is required, then the angle section can be inverted and, in so doing, the flange is moved away from the neutral axis (see figure 5.37). However, the designer could choose the following method, saving material and/or labour by disposing with the flanged angle section (see figure 5.38). To make full use of the scope of welded fabrications, the dimensions and shapes of components must be considered carefully.

The casting process has limitations, particularly with regard to varying wall thicknesses and closed box sections, but this is rarely the case with welded fabrications. It is possible, and it should always be one of the main considerations of the designer, to achieve a welded structure of low weight value. It may well be that a welded structure is less heavy than a steel or iron casting when completed, but the cost of plate actually used to achieve a low weight is of great economic importance. A plate layout drawing to afford minimum scrap material should be produced (see figure 5.39).

In the casting process, inspection holes, lightening holes, gusset and rib shapes are all automatically achieved. The choice of a welded fabrication entails flame-cutting or shearing of plates to provide these features. Therefore, it is possible, from the cost point of view, that the material used in fabrication is not reduced. Again, it is not always a practical proposition to design a structure without avoiding complicated features.

There are many instances where bosses are required to be welded to members, to accomodate bushes or bearings which, in turn, will support shafts. Although figure 5.40 names the cheapest and most expensive methods with regard to preparation, it is the accuracy of location and load-carrying capacity which are of prime importance.

Welding methods can only be applied efficiently if the workforce involved follow exactly the drawing issued to them. The engineering drawing is the specification which dictates the fabrication process, and it must contain all the information needed for correct manufacture. The following information must be provided, along with that shown in figure 5.41.

Plate and edge preparation
Size and type of welds
Welding sequences
Assembly procedure
Heat treatment
Inspection instructions

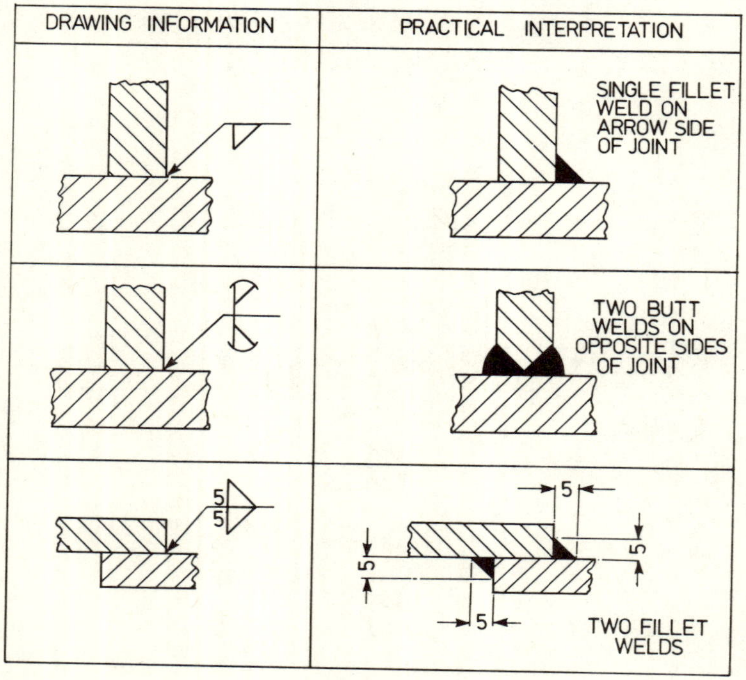

Figure 5.41

To emphasise this point, the plate layout ensures minimum plate requirements and provides economy of cutting. The size and type of weld should be to British Standard specification. It may be that only one sequence will ensure correct assembly of components or that distortion may occur during assembly, hence the need, in certain cases, to provide assembly procedures and welding sequences. If stress-relieving is required, then the heat treatment must be stated on the drawing.

As with all engineering design decisions, the design approach is all important (see figure 5.42). There is, of course, nothing to prevent the designer incorporating a steel casting into a fabricated structure. It is all a matter of the judgement of the designer, who may find that labour costs can be reduced by using additional material! The technical requirements of a design will affect design decisions, but the simplicity of fabricated design features influences the material consumption and, equally important, the labour costs.

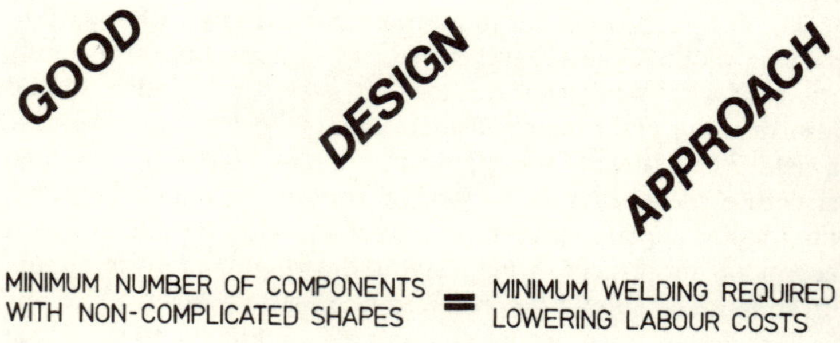

GOOD DESIGN APPROACH

MINIMUM NUMBER OF COMPONENTS WITH NON-COMPLICATED SHAPES **=** MINIMUM WELDING REQUIRED LOWERING LABOUR COSTS

Figure 5.42

The use of standard sections is of prime importance, and there is a large variety of shapes and thicknesses available. Corrugated panels, flats, rounds and rolled steel sections are typical (see figure 5.43). Consideration must also be given to the available shearing or guillotine capacity, which is less costly than flame-cutting or machining. Remember that the more complicated the structure, the more chance there is of excess scrap or offcut material.

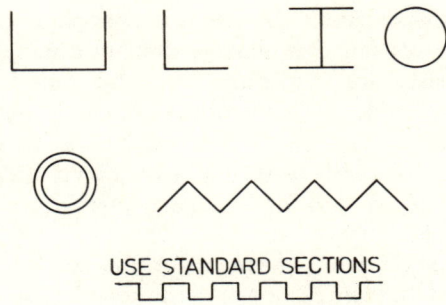

USE STANDARD SECTIONS

Figure 5.43

A Few Design Hints

The accompanying diagram (see figure 5.44) shows a stiffener fitting into a trough section. Note the clearance on the stiffener for the bend in the right-hand corner, and the clearance on the left-hand side to allow a continuous weld along the inside of the trough for the vertical member.

In certain assemblies a slight deviation from, for instance, the standard butt-weld specification can help assembly problems by allowing the vertical member to rest on the horizontal member. Again, a horizontal member is shown resting on a vertical plate, easing assembly problems, though consideration must be given to the size of fillet weld required (see figure 5.45).

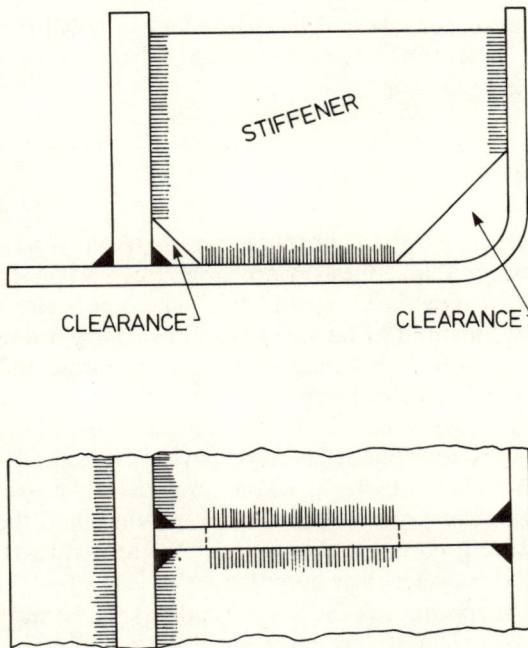

Figure 5.44

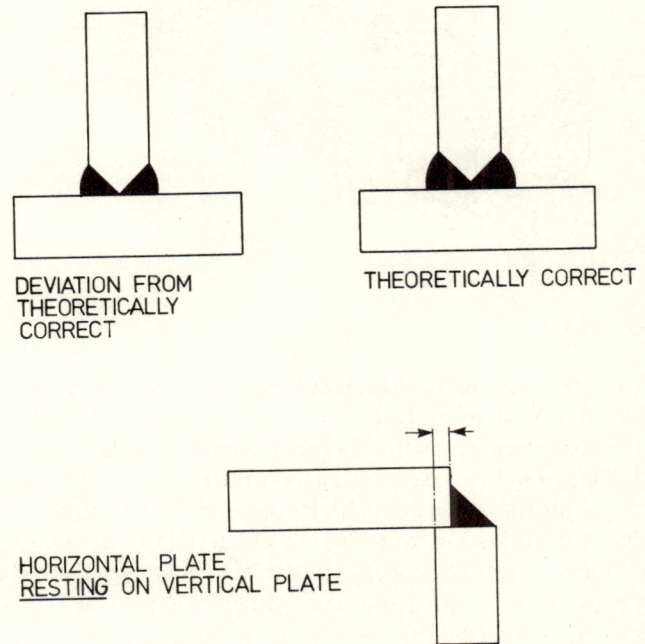

DEVIATION FROM THEORETICALLY CORRECT

THEORETICALLY CORRECT

HORIZONTAL PLATE RESTING ON VERTICAL PLATE

Figure 5.45

Finally, while a fabrication must be designed to meet service conditions, the problems of ease of fabrication and accessibility of welds may mean the use of a series of sub-assemblies. Such a decision may not only aid production of the complete fabrication, but can facilitate easier transportation and easier site erection.

Standard Components

Mention has been made in this chapter of the use of standard parts, in the form of rolled, pressed or drawn sections. The use of such items can reduce the cost of an assembly. However, there are other examples where the use of standard components can also eliminate design and fabrication in a manufacturing organisation (see figure 5.46).

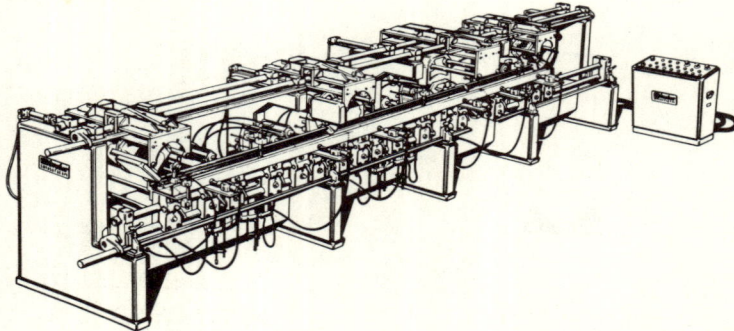

Figure 5.46

A new concept in machine construction is now being offered by Trend Elu Machinery Ltd of Borehamwood, Hertfordshire, the British counterpart of the German manufacturers, Eugen Lutz KG. This new Elu system is based on the selection of standard components to build machinery for a specific function, eliminating design and fabrication from basic raw materials. Basic components in this new modular concept consist of stands, bracketing, slides and clamps, all constructed from either cast iron or mild steel. Motor units are available for drilling, milling, routing and sawing operations, while pneumatic/hydraulic control equipment is provided to give part or full automation and to provide controlled feed-in to the work.

By using this assembly process it is possible to produce both simple and highly sophisticated machines for drilling and working light metal, cast iron, plastics and man-made boards. The electric motor units provide the high torque needed to obtain constant drilling and machining speeds. Running costs of these motors are approximately one-fifth of their pneumatic equivalents, while a further advantage is that they run particularly quietly even when mounted in batches.

The pneumatic/hydraulic rams for the self-feed motors are housed within the main casting of the motors themselves. Available with power ratings between 1 h.p. and 1.7 h.p. and with speed variations between 800 and 4000 r.p.m. obtainable through toothed gear and belt drives, they give drilling centres of less than 25 mm on multi-spindle heads and of 43 mm on single head units. Operation of all or selected power can be remotely controlled from a push-button console, which can activate not only the motors but also the pneumatic/hydraulic feed-in, clamping and optional automatic coolant spray.

ADHESIVES

Adhesive technology and products have, in recent years, developed to the extent that adhesives are now accepted by industry as providing a practical, reliable and in some cases the most economical method of fastening materials. They are used in the automotive, aircraft, building, electronics, furniture and footwear industries, and many others.

In recent years it has been estimated that approximately three out of every four manufacturers have found applications for adhesives. Undoubtedly in some instances they are the only fastening means possible, but their ease of use and the fact that manufacturing costs can be lowered are attractive reasons for industry to consider their adoption.

In some applications of adhesive bonding the cost and weight are lower than the equivalent mechanical joint, and fatigue dangers due to the drilling of joined parts are avoided. A basic consideration of types of adhesive is given in table 5.10.

Joining by adhesives is a simple process that does not call for

Table 5.10

Type	Materials	Description	General Remarks
Synthetic resin	Wood	Composed of a resin and a hardener which are mixed together. Resists water, acids, heat.	Joints should be cramped while the glue is setting.
Polyvinyl acetate (PVA)	Wood Paper Fabrics Card Leather	Supplied as a white liquid that is ready to use.	Joints must be cramped. Certain brands of PVA glues can be used for bonding new cement screeds to old concrete.
Epoxy resin	Metals Glass Rigid plastics Pottery Stone Wood Fabrics Leather	Supplied in two separate tubes containing a resin and a hardener, which are mixed in equal proportions. Joins non-porous materials. Resists boiling water, acids, oils, heat.	Hardens by chemical reaction unlike most adhesives which set by the evaporation of a solvent. Gentle heating accelerates setting.

Contact adhesive	Wood
	Rigid PVC
	  (including
	  plastic
	  laminate)
	Fabrics
	Rubber
	Leather
		A synthetic rubber and resin adhesive which is applied to both surfaces and allowed to become touch dry before the materials are brought into contact.	A strong bond is formed immediately. Ensure that materials are fixed correctly first time. Readjustment is possible only with a few brands. Not suitable for joining small areas subject to strain, such as furniture joints.

Rubber-base adhesive	Rubber Leather Fabrics Wood Paper Metals Carpets	A combination of minute particles of rubber suspended in water or other solvents. Resists water, oil.	Provides a strong, flexible bond.

Table 5.10 (*continued*)

Type	Materials	Description	General Remarks
Polystyrene cement	Polystyrene plastics (not expanded or foam polystyrene)	Composed of polystyrene dissolved in a solvent which fuses plastics.	
PVC	PVC plastics Vinyls Leather Rubber Fabrics	A combination of plastics and resins in a solvent. Resists water, oil.	
General purpose	Paper and card Wood Leather Canvas Fabrics PVC and rigid plastics Glass and metal Plaster	Most of these products, often called household adhesives, are based on cellulose, nitrile or polyurethane rubbers.	Although this type of adhesive sticks most materials commonly found in the home, it does not stick expanded polystyrene or polythene (the latter must be heat welded).
Cyanoacrylate	Metals Glass Most plastics Rubber Hardwoods	A vinyl monomer-base adhesive that sticks most non-absorbent materials in seconds. Most items can be handled in one to two minutes.	Should be spread thinly to ensure a strong bond. Surfaces must be smooth and free of hollows. Keep glue off hands.
Resorcinol resin	Wood Plastic laminate Asbestos	Two-part, fully weatherproof glue.	Mix together (five parts resin, one part hardener).

skilled labour. It is readily adaptable to automatic processing methods such as spraying, and large-scale use such as shoemaking and automobile production.

Adhesive joints are neat and unobtrusive—sometimes they are invisible. When adhesive bonding is used, smooth unbroken surfaces can be preserved—there is no need for bolts and rivets and the strength of the joint is spread equally over the whole surface. In addition, a layer of adhesive can act as a perfect seal against liquids or gases; many adhesives are resistant to a wide range of chemicals and solvents.

The basic function of an adhesive is to cement two or more surfaces together. The result is a sandwich of adhesive between the surfaces. To cement surfaces together, the adhesive must wet the surfaces and then harden (see figure 5.47).

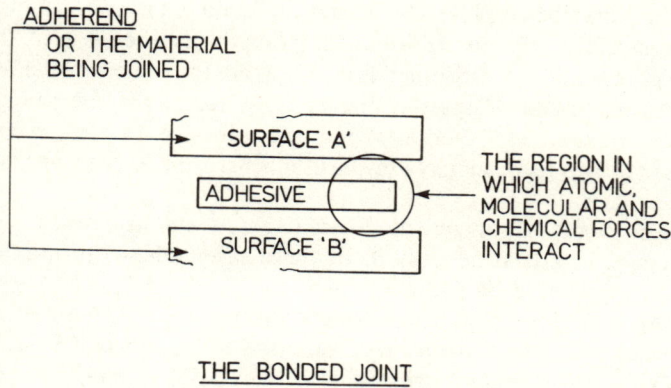

ADHEREND
OR THE MATERIAL
BEING JOINED

SURFACE 'A'

ADHESIVE

SURFACE 'B'

THE REGION IN
WHICH ATOMIC,
MOLECULAR AND
CHEMICAL FORCES
INTERACT

THE BONDED JOINT

Figure 5.47

THE LOAD AT WHICH A LAPPED JOINT
WILL FAIL IS PROPORTIONAL TO THE
WIDTH 'W' BUT NOT PROPORTIONAL
TO THE LENGTH 'L'

W

L

Figure 5.48

Thermohardening adhesives contain epoxy resin. Heat and pressure are required, and the joint sets hard. The joint cannot be taken apart after setting. *Thermoplastic* adhesives contain polyvinyl resins. Heat is not required but pressure is needed. The joint can be melted by heating or softened with solvents. The impact type of adhesive is thermoplastic, and thermoplastic adhesives are generally less rigid and weaker than the thermosetting type.

There is a very wide variety of physical and chemical types available, from which a suitable product can be chosen to solve most bonding problems. Bonding is frequently the best way of joining dissimilar materials. Joints of great rigidity can be made— or the adhesive layer can be flexible to absorb vibration or to take up stresses caused by different rates of expansion (see figure 5.48).

Sometimes a bond fails owing to a weakness in the material of one or other of the joint components, either because of poor adhesive contact with the surface of the component or incompatibility between the adhesive and the component. The adhesive must wet the joint surfaces to perform efficiently, and it can only do this if the surfaces are perfectly clean and correctly prepared.

It is vitally important that all contaminants such as water, grease, oil, dust, rust, etc., should be removed before adhesives are used.

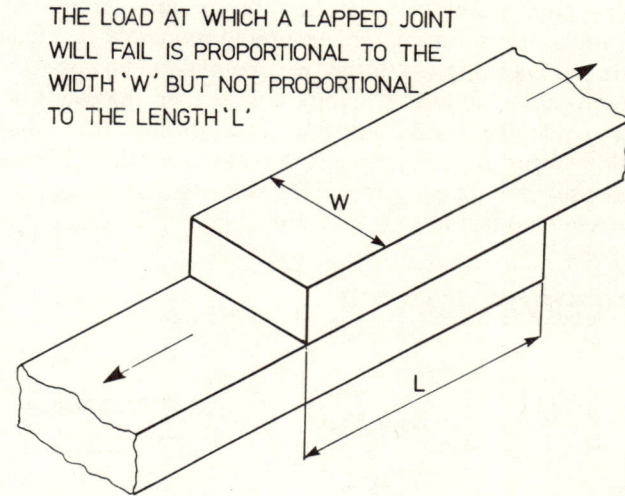

In general, the best surface treatment is to degrease, abrade and, if possible, degrease again to remove abraded dust (see figure 5.49).

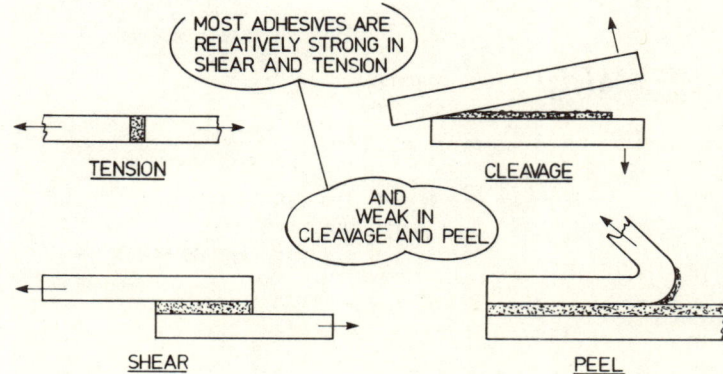

MOST ADHESIVES ARE
RELATIVELY STRONG IN
SHEAR AND TENSION

AND
WEAK IN
CLEAVAGE AND PEEL

TENSION

CLEAVAGE

SHEAR

PEEL

Figure 5.49

Inadequate bonding between the adhesive films can usually be traced to a defect during application. Frequently, insufficient adhesive has been applied to the two materials to be bonded. If a continuous adhesive film is plainly visible, then the cause of the trouble could be inadequate pressure during the bonding operation or too long a drying time. A point to watch for here is the possible presence of high spots on the surfaces which may prevent good overall contact (see figure 5.50).

ADHESIVE JOINTS

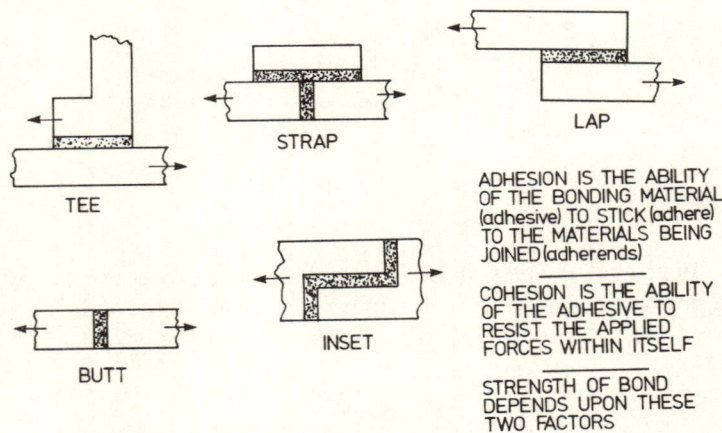

TEE

STRAP

LAP

BUTT

INSET

ADHESION IS THE ABILITY OF THE BONDING MATERIAL (adhesive) TO STICK (adhere) TO THE MATERIALS BEING JOINED (adherends)

COHESION IS THE ABILITY OF THE ADHESIVE TO RESIST THE APPLIED FORCES WITHIN ITSELF

STRENGTH OF BOND DEPENDS UPON THESE TWO FACTORS

BEWARE! SKIN IRRITATION
TOXIC VAPOURS
INFLAMMABILITY

USE GLOVES - BARRIER CREAM
VENTILATION
NO NAKED FLAME - AND DO NOT SMOKE!!

Figure 5.50

The right tack retention time is also very important and this depends on the technique of application. The 'two-way-dry'

method may be employed, in which the bond is made during the tack life of the two films of adhesive. When this method is used, the two components to be joined are kept apart for some time before the bond is made. Using the 'one-way-wet' technique, the bond is made immediately after applying the adhesive to one of the surfaces. This method may only be used where one or both surfaces are porous.

In some cases it may be necessary to reactivate the dry adhesive by solvent or heat activation. In these cases adequate time for activation should be allowed.

The considerations which follow outline the selecting of an adhesive and the preparation of materials to be joined. The chart showing characteristics, uses, etc. (see figure 5.51), is reproduced by kind permission of Bostik Ltd, of Leicester, and the large range of products supplied by this particular company is some measure of the choice afforded to both the do-it-yourself enthusiast, craftsman and industrial manufacturing unit.

Selecting the Correct Adhesive

For any one particular bonding application there is one adhesive formulation that is best suited to do the job. Although the final selection may well require a detailed study of the bonding problem, there are four basic factors that must be considered in the initial selection process. Choosing an adhesive to suit a given set of factory conditions can have far-reaching effects, because the properties of the adhesive can influence output rates, consumption and performance.

Materials

The adhesive that is selected must be the one that will adhere effectively to the materials which require bonding. The chemical nature of the adhesive and also of the substrates, together with the surface condition of the substrates, must be considered.

BOSTIK	DESCRIPTION	CHARACTERISTICS	USES	BASE	COLOUR
BCA3	Synthetic Rubber/Resin Adhesive	Quick drying, long tack life. Highly flammable.	Specially formulated for bonding decorative laminated plastics to wood, hardboard etc.	Polychloroprene	Buff
"C" Adhesive	Reclaim Rubber/ Resin Adhesive	Excellent ageing properties. Water resistant. Highly flammable.	Heavy duty, general purpose adhesive. Will bond most porous and non-porous materials, e.g., rubber, linoleum, cork, felt etc. to wood, metal, etc.	Reclaim	Black
Clear Adhesive	General Purpose Synthetic Rubber/ Resin Adhesive	Dried film is virtually colourless and non-staining. Highly flammable.	Provides good bonds to PVC, Nitrile rubber, Fibreglass (GRP), P/U and PVC Foam, Metal, Wood, Textiles etc.	Nitrile	Clear
183 C	Vapour Seal Adhesive	Good residual tack and high resistance to water vapour. Non-flammable.	Bonds expanded polystyrene, expanded PVC, cork etc. Used on refrigerated transport, insulated containers, ice cream cabinets, sectional cold rooms.	Bitumen Rubber/Emulsion	Black
247	Bostik Solbit	Filled blended bitumen, anti-corrosive and high impact resistance. Sound deadening and anti-vibration. Highly flammable.	Brush or spray application for underbody coating of vehicles.	Bitumen	Black
281	Bostik Colset	Provides excellent resistance to water, humidity and moisture vapour. Resistant to fire spread when dry. Flammable.	Gives long term protection to treated roofs. Weatherproof anti-corrosive finish to flat and corrugated metal, felt, pitched or flat asbestos roofs.	Bitumen	Black
590	Chassis Black	Gives good protection in adverse conditions. Excellent resistance to salt spray. Very fast drying. Highly flammable.	Chassis black paint giving good preservative and anti-corrosion protection to all types of metalwork.	Asphalt/Resin	Black
6	Reclaim Rubber/Resin Sealing Compound	General purpose sealing compound supplied in tubes only. Excellent weather resistance. Non-flammable.	General purpose sealing compound for glass, metal etc.	Reclaim	Black
772 771 in tubes	Synthetic Rubber/ Resin Adhesive and Sealing Compound	General purpose adhesive and sealing compound where resistance to oil and petrol is required. Highly flammable.	High strength bonds to metal, glass etc.	Nitrile	Blue-Black
1297	Rubber/Resin Adhesive	Excellent water resistance. Has temperature range of −17°C to +100°C. Highly flammable.	Bonds to variety of surfaces, i.e., rubber, cork, linoleum, felt, wood, metal.	Natural Rubber	White
1311	Panel Adhesive	One way bonding technique providing good initial adhesion. High viscosity, water resistant. Highly flammable.	Provides fast fixing method for plasterboard, hardboard. Particularly suitable for application to irregular surfaces.	Natural Rubber	White
1530	Sealing Compound	Odourless sealing compound, good resistance to water and humidity. Non-flow with good ageing properties. Will not contaminate foodstuffs. Non-flammable.	Specially developed for sealing operations in cold room construction and refrigerator manufacture.	P.I.B.	Off-White
2000	Two part Epoxy Resin Adhesive	Solvent free, flowable, cold curing. Does not shrink and has good water and chemical resistance. Non-flammable.	Provides high bond strength to wood, glass, pottery, steel and aluminium. Particularly useful for cutlery hafting and similar operations.	Epoxy	Brown
2024	Two part, cold curing Epoxy/Polysulphide Adhesive and Sealant	Weatherproof, dustproof, waterproof. Resistant to oil and petrol. Does not shrink or slump. Provides hard, slightly flexible seal. Non-flammable.	Provides good bonds to glass, metal, concrete and wood. Used extensively as sealant in manufacture of double glazed units.	Epoxy-polysulphide	Dark Grey
2117	Two part Polysulphide Caulking compound	Provides resilient seal, high tensile strength and elongation properties after prolonged exposure to weather. Highly resistant to oil, grease, fresh and salt water. Highly flammable.	Specially developed for permanent sealing of deck joints on all types of sea-going, and small craft. Good adhesion to wood and metal.	Polysulphide	Black
2135 2137 2138	Two part Polysulphide Sealants	Cold curing. Durable and weatherproof. Will accept considerable movement. Meet requirements of BS 4254/1967 ASA 1161/1960 and American Fed. Specs. TTS 00227 A & B. 100% solids. Non-flammable.	Sealing expansion joints in walls, joints between cladding panels, curtain walling, window frames, glazing etc.	Polysulphide	2135 Black 2137 Grey 2138 White

Figure 5.51

BOSTIK	DESCRIPTION	CHARACTERISTICS	USES	BASE	COLOUR
2221 2222 2223 2224	Two part Viton synthetic sealants	Exceptional resistance to high temperatures, fuels, chemicals and abrasion. Highly flammable.	Specially developed as integral fuel tank and pressure cabin sealants on supersonic aircraft. Also have application in cars, electronics and atomic energy where high temperature resistance, flexibility and fuel resistance are essential.	Viton Viton Viton Viton	2221 Black 2223 Black 2222 White 2224 White
2402	Two part cold curing Synthetic Rubber/Resin Adhesive	High strength, good initial tack, excellent resistance to water. Highly flammable. Use Bostik 9252 Primer for best adhesion to metal, glass & masonry.	Provides strong bonds to natural, and synthetic rubbers, rigid PVC, wood, leather, fabric, metal, glass and masonry.	Polychloroprene	Buff
3206	Synthetic Rubber/Resin Adhesive	General purpose, quick drying, easy application. Highly flammable.	Provides good bonds to wood, canvas, leather, nylon and PU coated fabrics. Particularly good for plasticised and unplasticised PVC and Decorative Plastics Laminates.	Polyurethane	Clear Amber
4141 4142	Synthetic Polymer Adhesives	4141 – Med. viscosity. 4142 – Low viscosity. Water based and good in roller coating applications. Water resistant when dry. Non-flammable.	Suitable for bonding PVC, wood, metal, paper, hardboard etc.	Synthetic Resin Emulsion	White, colourless when dry
6360 6365	Thermogrip Hot Melt Adhesive	General purpose hot melt adhesive for use through Thermogrip Plug Gun. Non-flammable.	General purpose wood/wood, expanded polystyrene, paper, board etc.	E.V.A.	Cream
5653 Prestik	Extruded Sealing Strip	Good ageing properties on prolonged exposure. Remains plastic between extremes of heat and cold. Water resistant. Non-flammable.	Joining and bedding concrete blocks and panels. Fixing roof lights and ventilators. Sealing joints in sectional building.	Oil/Asbestos	Cream
5686 Prestik	Extruded Sealing Strip	Resistant to water, salt water. vegetable oil and coal gas. Satisfactory sealant for temperatures up to 250°C. Non-flammable. Non-Toxic.	Specially formulated for sealing joints in gas and electric cookers. Satisfactory at high operating temperature.	Oil/Asbestos	Dark Grey
5703 Prestik	Extruded Sealing Strip	Resistant to water, salt water and vegetable oils. Excellent low temperature resistance. Odourless – will not contaminate foodstuffs. Non-flammable.	Specially formulated for use as a refrigerator and cold room sealant.	P.I.B.	White
5913 Prestik	Extruded Sealing Strip	Excellent ageing when exposed to the elements. Weatherproof, dustproof, flexible. Non-flammable.	General purpose sealing of sectional buildings, roof lights, bead glazing for glass up to 2 feet square.	Butyl Rubber	White
5925	Extruded Sealing Strip	Non-flammable and non-toxic, weatherproof, dustproof, will accommodate considerable relative movement.	Bead glazing, sealing panels and joints during erection. Curtain wall sealing, fixing roof lights and ventilators.	Butyl Rubber	Buff
9015	Asphalt Tile Adhesive	Cold setting. Flammable. Water resistant.	Fixing thermoplastic asphalt tiles and vinyl/asbestos flooring tiles to concrete and other sub-floors.	Bituminous	Black
9100	Film Adhesive	Dried non-flammable film adhesive on non-woven carrier. Solvent or heat re-activated.	Bonding name plates to die-cast aluminium. Will also bond to steel, ABS, rigid and plasticised PVC.	Nitrile	Translucent film on white carrier
9105	Hot Melt Film Adhesive	Heat activated, non-flammable, instant bonding. Resistant to hot water (90°C) and perchlorethylene.	Specially designed for bonding fabrics, sealing seams in waterproof garments etc. Also provides good bonds to aluminium foil, ABS, PVC, steel, leather, wood and paper.	Polyester	White Translucent
9410	Thermogrip Hot Melt Adhesive	½″ diameter grooved rod form adhesive for use in Thermogrip Applicators. Non-flammable. 9381/9382—Granulated Form	High speed bonding in packaging industry. Bonds to polyethylene coated board. Withstands deep freeze.	Polyethylene	Straw
6361 6366	Thermogrip Hot Melt Adhesive	General purpose hot melt adhesive for use through Thermogrip 260 Electric Glue Gun. Non-flammable.	General purpose wood/wood, expanded polystyrene, paper, board etc.	E.V.A.	Cream
—	Sheet Form Sound Deadener Pads	Heat fusible on to metal. Will withstand subsequent stoving operation. Excellent sound deadening and anti-vibration properties. Non-flammable.	Specially developed for the auto industry, but can be used on any metal fabrication where heat fusing is possible.	Bitumen/Rubber Blend	Black
—	Sheet Form Sound Deadener Pads	Pressure-sensitive pads which can be applied to any clean metal surface. Excellent sound deadening and anti-vibration properties. Non-flammable.	Specially developed for the auto industry, but can be used on any metal fabrication.	Bitumen/Rubber Blend	Black

Figure 5.51 (*continued*)

Service Conditions

The correct adhesive must be able to withstand extremes of temperature, impact, pressure, movement, corrosive elements, environmental and other factors which may be encountered during the service life of the bonded product.

Production

The correct adhesive properties must relate closely to the production requirements

(1) allowable time between application and bonding
(2) desired tack, drying or curing time
(3) efficiency of application
(4) assembly limitations
(5) plant environmental conditions.

Costs

The cost per container of adhesive is not the only cost consideration when selecting an adhesive. Equally important are the following

(1) solids content coverage
(2) the reduction of rejects
(3) efficiency of application
(4) the value of the added quality in the finished product.

Proper Surface Preparation

For optimum bonding results it is essential that clean surfaces be provided prior to the application of adhesive. The surfaces should be free from contaminants such as rust, moisture, dust, oils and mould release agents. The strength of bond required and economic limitations directly affect the amount of surface preparation but, generally speaking, the more thorough the preparation, the stronger the adhesion. There are three basic cleaning techniques which can be applied.

Chemical

Etching or chemical cleaning is a popular method of metal surface preparation. This method ensures maximum adhesion.

Abrasive

Sand-blasting, wire-brushing or sandpaper is adequate when the surfaces to be bonded are of sufficient thickness to prevent distortion by such cleaning methods. Abrasive cleaning is more economical and can be completed more quickly, together with a rougher bonding surface, than the chemical or degreasing methods.

Degreasing

This method is necessary for the removal of residual contaminants such as oil, grease, etc., from the surfaces to be joined.

The Use of Solvents

Solvents perform important functions in adhesive and sealant technology. Solvents are liquids that are capable of dissolving other substances, and a solvent's power is measured specifically by what it can dissolve. In adhesive technology, solvents are used to convert the solid-base materials into liquid form so that they can be easily applied. Also, the solvent's rate of evaporation has a direct effect on the drying and curing characteristics, that is, the solvents which have slower evaporation rates are used to produce adhesives or sealants with longer open times which will dry or cure more slowly. A further role of the solvent is the effect it has on the substrates to be bonded. The ideal solvent will 'wet' thoroughly, but will not damage the substrates, so that optimum adhesion is ensured.

6 The Product and Factors Affecting Product Appearance

Regardless of the type of product under consideration, there are certain factors which affect the product appearance. Domestic appliances, cutlery, clothes, furniture and cars all require the skill and flair of the trained designer, to meet the need for order, variety, proportion and symmetry in the finished article. It is necessary therefore, to understand the meaning of these terms.

Order In the briefest way possible, one could say that order means 'tidiness'. It is a condition in which every part or unit is in its right place. A simple example is that of alphabetical order.

Variety Here is a word with meanings that can apply in many design situations.

(1) Diversity, absence of uniformity.
(2) An accumulation of unlike things.
(3) A class of things (or specimen of it), differing in some qualities from the rest of the larger class which includes it. For instance, there are many types of fastener, used in engineering assemblies, generally referred to as bolts or screws. Within the large class of fasteners, there are countersunk, hexagonal and cheese head screws. The differing quality, affording variety, is the type of screw head.

Proportion The relationship in size between connected things or parts. Consider the proportion of the teacup, cup handle and matching saucer, or the diameter of bolt shank relative to the bolt head. Correct proportion can therefore lead to the definition of symmetry.

Symmetry This is the pleasing harmony or effect resulting from the correct proportioning of parts (see figure 6.1).

Each of us has differing views regarding, for instance, the style of car body which we prefer. The body design team attempt to achieve order, variety, proportion and symmetry, with the object of attracting a large market. The same reasoning applies to audio equipment, freezer units, washing machines, electronic calculators and cameras.

All products necessitate company expenditure and investment, which it is hoped can be converted into profit by persuading the customer to make a purchase.

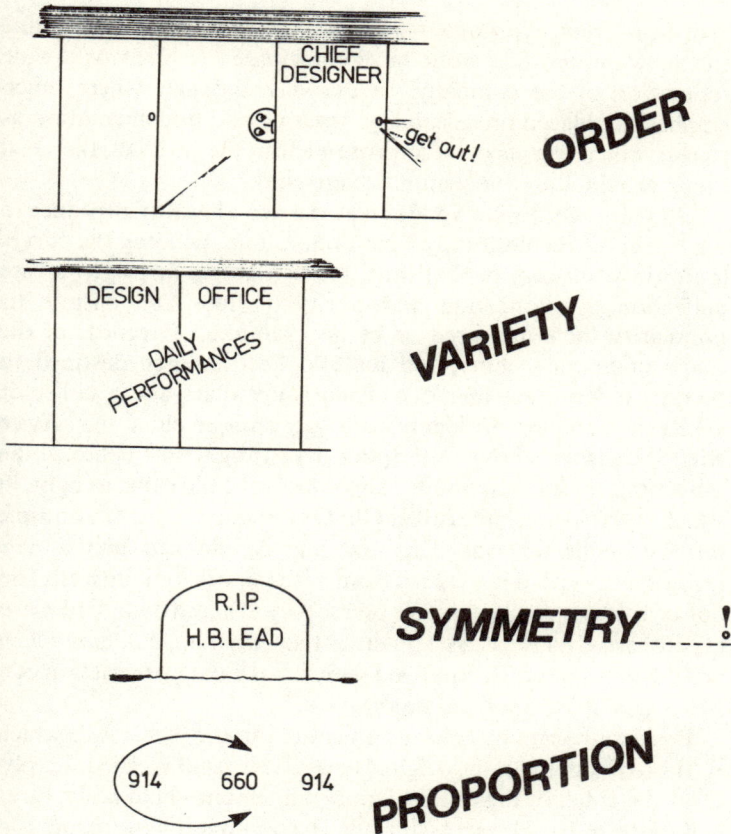

Figure 6.1

is necessary, whereas beauty or good taste need not be considered. A pump in an underground sewage system or the shovel attachment on earth-moving equipment neither receive nor require aesthetic considerations. However, there are many products in everyday use that require combined aesthetical and functional qualities.

What is beauty? What is good taste? Such things are in the eye and mind of the beholder, and can be the dilemma of the designer. Knives, forks and spoons are in constant daily use, as also are wristwatches, cups and saucers, teapots and cars. In good taste or otherwise, Mr and Ms Everyone spend millions of pounds on what they require, and what they require commences life in the design office. Designers are merely human and their taste should not override that of the public at large.

For the moment let us concern ourselves with products that have both aesthetic and functional qualities, that require symmetry in the finished form and that are so often taken for granted. The design student should consider the materials from which the products are made, and think for a while about the tools, equipment and machinery required for manufacture. Also consider the skill of the craftsmen and the final finish of the product.

Saleability, advertising and market research have not yet been mentioned, but what has been said is that the designer is only a part of the successful product story. The designer is a person with a role—a fact which he must accept with modesty and yet at the same time also accept responsibility for design decisions.

An original, unique and distinctive product is the combination knife, fork and spoon, manufactured by Viners Ltd of Sheffield. It is called Splayd and is a revolutionary, all-purpose eating item manufactured from 18–8 stainless steel. The aesthetic features are the satin-finished shank and highly mirror-polished bowl. Functionally, the features are such that they prompt the comment: 'Why didn't I think of that!' It can be used like a knife, by both right- and left-handed people, a sharp edge being provided on each side of the bowl. It scoops like a spoon and efficiently spears food on the four prongs. The object of the Splayd is to transfer food to the mouth, without spillage, and it is particularly useful at functions such as stand-up buffets. A point of particular interest to

There are many ways of comparing products, and the words 'aesthetic' and 'functional' are frequently mentioned with reference to product design. Aesthetic means being concerned with or appreciation of the beautiful, or more simply, good taste. A product is said to be functional if it achieves the task or work for which it was originally designed. Many engineering components and assemblies are functional without having aesthetic properties. Indeed, there are many instances where one could say that function

those who are keen on labour-saving ideas is that, because only one item instead of three is involved, the use of a Splayd makes for less washing-up and less pre-meal preparation (see figures 6.2 and 6.3).

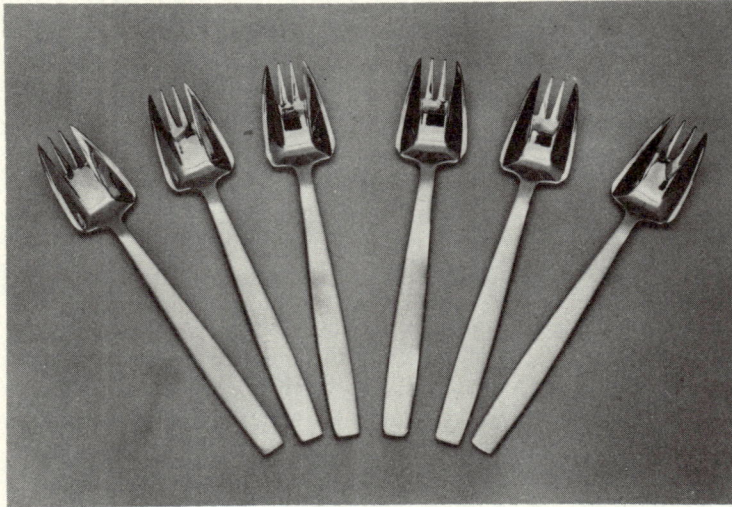

Figure 6.2

Figure 6.3

When defining aesthetic as being concerned with, or having an appreciation of, the beautiful, then surely one of the best examples of this definition is the Oval range of tableware, by Spode Ltd of Stoke-on-Trent. Top-quality products in fine bone china, fine earthenware and fine stone have contributed to the world-wide reputation of the company. In this day and age, where much emphasis is placed on saleability, research and documentation as factors which can effect an improvement in the product, the Oval range provides an interesting background.

175 years ago, Josiah Spode created a new shape to introduce to the world his revolutionary Fine Bone China. Echoing the curves found in silverware of that time, his new teaset displayed a rare perfection of proportion and purity of line. Yet, despite its popularity there appeared to be no examples or records of the shape of Spode's china, and his Old Oval seemed destined to become a forgotten memory. Then some years ago a collector asked the company to identify a lovely antique china tea service which he suspected might be Spode. By strange coincidence, at the same time, an original Spode shape-book of 1820 came to light, in which there was a shape called Old Oval that matched the antique set. The collector was delighted and Spode Ltd had a new inspiration. With this shape in front of them, still looking fresh and full of sales appeal after 175 years, they realised that Old Oval encompassed all the classic qualities they insist on; qualities which will never go out of fashion and which make the introduction of a new shape from Spode a major event.

The design team has created an entire range of tableware which is true to the shape of the original teaset. The result is called, simply, Oval. To enhance the beautiful pure white bone china body, three new patterns have been originated. The photograph (see figure 6.4) shows the Country Lane pattern. Although the teapot is based on a silver design of the Georgian period, and nearly all the other items have been designed to fit in with this, at the same time they meet the functional demands of everyday living.

Electrolux Ltd of Luton, Bedfordshire have a philosophy: 'Good Enough' isn't good enough for us. They are the manufacturers of a large range of domestic equipment, including vacuum cleaners, floor polishers, refrigerators, freezers, electric radiators, dishwashers and cooker hoods. Their service organisation is nationwide. Two of their latest models are illustrated (see figures 6.5 and 6.6).

Figure 6.4

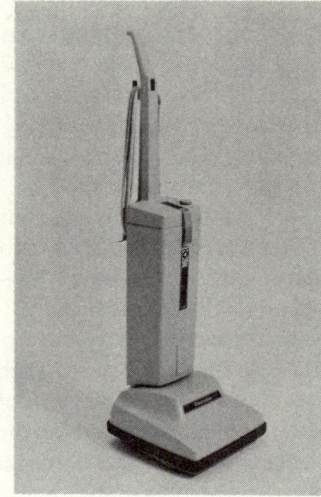

Figure 6.5

The Electrolux Twin 502 upright cleaner glides on four wheels, and the handle lowers almost horizontally for cleaning under low furniture. High-suction power lifts out dirt, a double row of brushes grooms carpets. The cleaning head adjusts automatically for different thicknesses of carpet—and hard floors. The hose simply plugs in, to give powerful suction for all-over cleaning. Its accessory kit consists of hose, extension tube, 'flip-over' carpet/floor tool, dual-action dusting tool and crevice nozzle. It is light and manoeuvrable, weighing only 13 lb, and it is quiet too. Simple control reduces suction for cleaning lightweight and thin, rubber-backed carpets. A whistle signal warns when the dust bag needs changing. Hygienic disposable paper dust bags (can be re-used), simply lift out and nothing gets spilled. Returned air is triple-filtered. It has a quick-release flex holder with a handy thumb switch. Its colour scheme is avocado green with dark brown trim and its aesthetic qualities are clear to any observer (see figure 6.5).

The next illustration (see figure 6.6) shows the new Electrolux Automatic 345 cylinder cleaner. It has an automatic cleaning head that lifts its brushes for carpets and lowers them for hard floors.

Figure 6.6

The powerful suction maintains high efficiency through every cleaning job. Owing to its automatic bag-change control, when the dust bag needs changing the Automatic 345 stops, a new dust bag is inserted and the motor re-starts. An ordinary cleaner with a full dust bag can go merrily on, picking up next to nothing. The quick-change, self-seal, throw-away dust bags have a unique rubber seal, which closes automatically for easy, hygienic changing. A 'mains-on' indicator light shows when power is getting to the cleaner and goes out when motor starts. There is an automatic flex re-wind, the flex tucking away tidily inside the cleaner.

The Automatic 345 has the following simple, multi-purpose tools: the automatic cleaning head; a dual-action dusting brush/nozzle for stairs and furnishings; a crevice nozzle with a detachable brush head for awkward places; fingertip suction control; a rotating hose allowing free movement with tools; an air filter and diffuser. Its colour scheme is two rich shades of brown.

Once again the aesthetic qualities are evident. The product is sleek, streamlined and handsome. The advertising literature issued by Electrolux Ltd makes the suggestion that the prospective purchaser should choose which vacuum cleaner 'you would most like to live with', and continues by adding, 'after all it is going to be with you for a very long time'.

The engineering design student, or any design student, should remember that the quality and reliability of a product will keep the customer happy for a very long time.

Clean modern styling is illustrated in the cars produced by Vauxhall Motors Ltd of Luton (see figures 6.7 to 6.10). Apart from the obvious proportion and symmetry, providing the aesthetic features, the functional attributes cater for every type of motorist. Vauxhall claim that their large range of cars is built for the teeth-rattling door slammer, the neck-jerking clutch prodder, the every-pothole-seeker, the stationary wheel-steerer; the ratchet-clicking handbrake puller, and a few others besides!

The examples of design which have been considered include the use of *styling* which is one of the aspects of *fashion*. Cars and clothes are examples which provide never-ending stylistic changes, which create desire in the consumer to replace them well before the end of the useful or working life of the product.

Planned obsolescence is a term often used to describe how a predetermined length of working life can be designed into a product. Ideally, the length of the life cycle coincides with stylistic change. In such cases the designer cleverly conceals the factor of built-in obsolescence.

Figure 6.7

The Vauxhall Magnums

Figure 6.8

Figure 6.9

The Vauxhall Cavaliers

Figure 6.10

not just a matter of what is easiest to manufacture, but rather what the consumer is willing to buy. The creativeness of the designer must be orientated to people and their needs, rather than to products. This leads us to the next consideration, that saleability and research can effect an improvement in the product.

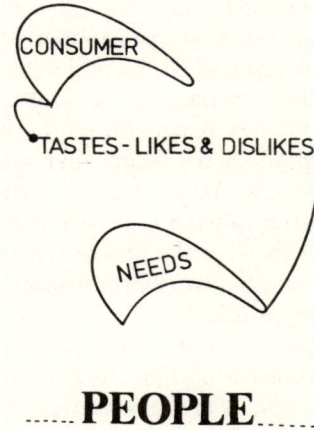

Figure 6.11

It is perhaps now that the function of the designer, as part of a large team whose ultimate aim is to provide for consumer needs, is beginning to emerge. It is sometimes easy to lose sight of the fact that we are all consumers from the cradle to the grave, and that each individual has specific needs (see figure 6.11).

One cannot deny that pressure on the individual is created by marketing methods, to the extent that the consumer lives in two worlds. There is the world of the real product and the world of the product as highlighted by marketing, advertising and selling techniques (see figure 6.12).

The functions of marketing and design must be in unison. It is

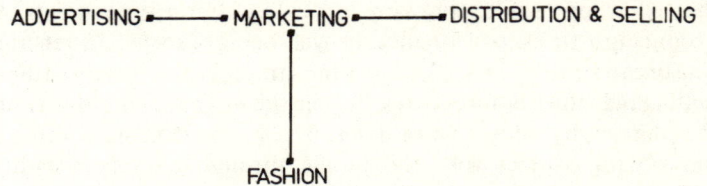

Figure 6.12

MARKETING A PRODUCT: SOME BASIC CONSIDERATIONS

Obtain some money—manufacture some products—sell the products—hope for profit. This concept meant that marketing could be defined as selling an existing product at a rate that would provide continuous work for production facilities. In the competitive world of today, the basic idea of 'factory produce something, marketing sell it' is open to many hazards. It will be noticed that the customer and choice of purchase have not, so far, been mentioned. The customer is the source of profit for the manufacturer, and for this reason alone the modern marketing concept concerns more than making and selling. The modern manufacturing unit must be customer orientated. As already stated, it is not a matter of what is easiest and most profitable to manufacture, but rather what the prospective customer or user is willing to purchase. In other words, the wants and needs of people have priority over the product.

In terms of design discipline and eventual company profit, much is to be gained by considering first the market and secondly the production of the article. A manufacturing company succeeds by searching for ways to satisfy human needs. These needs, as already illustrated, combine functional and aesthetic characteristics, but the user also requires convenience, service and satisfaction (see figure 6.13).

Although marketing, in the broadest sense, can mean bringing together the buyer and the seller, the overall social interaction is that of a profitable company providing full employment by producing articles that enhance the well-being of society in general. The broad strategy of marketing is the same for cars, cutlery, lathes, milling machines or fuse-boxes, but the the methods employed and the philosophy adopted can differ. The manufacturing company can obtain contact with the public through television, radio, advertising and personal calls. The mass media methods are used for marketing cars, freezer units, double-glazing units and similar products. Such items are manufactured in large quantities, stored and distributed to retailers. However, engineering products such as lathes, drills, reamers, measuring equipment, etc., are usually seen at exhibitions, advertised in trade journals and form part of a

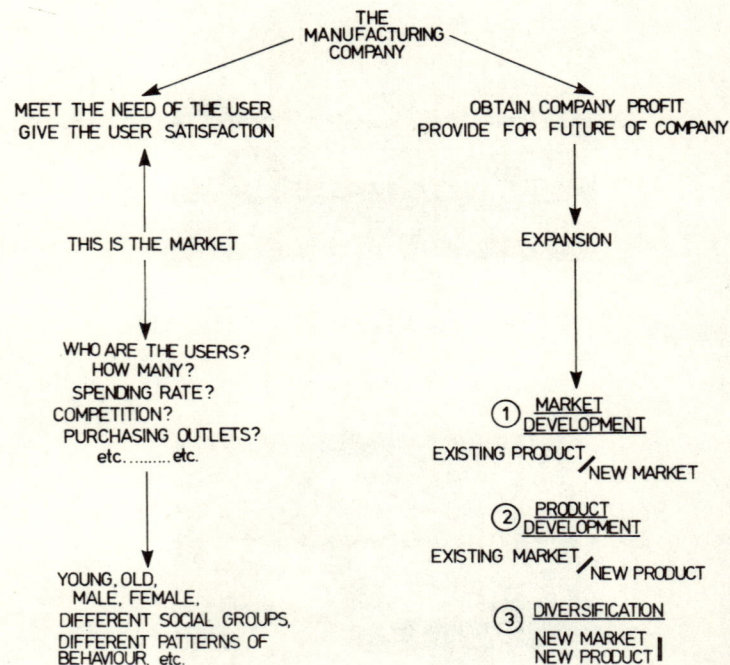

THE MARKET PLACE

Figure 6.13

smaller and more specialised market. A specific customer requirement is the only outlet for the sale of the product.

It will be appreciated that many attachments can be purchased for fitment to standard lathes, milling machines and other machine tools, both to aid production and to obtain greater product accuracy. Jigs and fixtures are further examples of production aids, and these are often designed and manufactured by specialist firms, who obtain an order from a particular client which necessitates close inter-company liaison. Manufacturers of varied specialist equipment or attachments usually have a standard range that can be modified to suit particular customer requirements, and the supplier will, if necessary, design and manufacture special 'one-off' products. The early stage of events in a user situation, requiring, for instance, a special attachment, can be as shown (see figure 6.14).

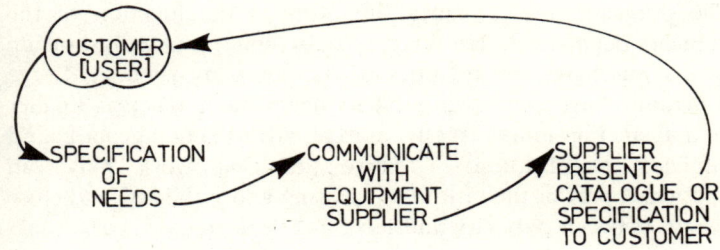

— SHOPPING AROUND —

Figure 6.14

Letters, telegrams and telephone calls play an important part in the early stages of buying and selling, but the time will soon arrive when there will be a need for the representatives of the user and the supplier to meet for detailed discussions. Let us look at the functions of the various personnel and departments for the marketing of a product in an engineering organisation.

Technical Representative (Outside Sales Engineer)

The duties of the technical representative are to interpret customer requirements. In doing so, the application needs will be discussed which may, in certain cases, benefit the customer financially, that is, the problem is perhaps not so complicated as the client envisaged. However, the customer may have underestimated the problem and will then have to be informed of special requirements. In either case the representative must help the customer to make a decision. He must have a good engineering background and be able not only to converse with a client, but also be able to transmit the customer's requirements back to supporting staff for action to be taken. The presentation of information in a clear manner, a logical approach for obtaining information from the customer and certain psychological skills are the essential requirements of such a person in the marketing team.

Estimating Engineer

The estimating engineer will prepare a price for the customer's requirements from the information received from the technical representative. A technical estimate will then be prepared and the customer will be informed. Specialised knowledge in the company's product with a previously taken apprenticeship and draughting/design experience are the requirements of an estimating engineer. The successful preparation of specifications and tenders will also require a knowledge of the commercial functions of buying and selling. As with the technical representative, the expertise of customer liaison is essential.

Contracts Engineer

On receipt of a customer's order, the contracts engineer will check it against the original quotation supplied, negotiate any differences that may be found and then start the procedure of manufacture for the order. A good engineering background, experience with company products and knowledge of company law are basic requirements for the effective functioning of this member of the team.

Service Engineer

The service engineer should obviously have a good engineering knowledge, with particular reference to the company's product, together with the ability of being able to work independently of others and to make decisions. A feedback of information to the technical department is an invaluable part of the task, both in the form of modifications that could be incorporated in future designs or justifiable criticisms that may be discovered in the fitting and commissioning of the product for the customer.

Others Involved between Buyer and Seller

Information can only be of value if it is presented in true form and made available for future reference. In a manufacturing organisation the information on a drawing, which should be filed correctly in a cabinet for easy future reference, is as important as a correctly prepared and correctly filed company balance sheet. This is to say that the often unrecognised person who has the task of filing drawings is as necessary as the managing director. Each has his own task. Each is dealing with information that helps to form the important communication chain which is the lifeline of industry and commerce. Here then are further marketing functions which form the blend between conception of design and the sale of the product.

Customer Liaison and Order Progressing

The collecting of customers' information, the feed-through of this information within the manufacturing company and the feedback of information to the client are the tasks performed by this section of the support team.

Export Department

Distance alone is not the only difficulty encountered when dealing with overseas customers. There are the problems of language and also foreign technical standards which must be considered in depth at both the quotation and acceptance-of-order stages. Within the manufacturing company it is essential that the export department supplies the various technical departments with the terminology and specifications of overseas countries.

Technical Publications

New products or technical improvement to existing products make it essential that technical literature is updated. Various commercial and legal aspects must be considered in the preparation of literature, and liaison with other internal technical departments is critical for the production of operating and maintenance manuals.

Publicity

The publication of technical literature is the function of the publicity department. However, it is also concerned with showing to the purchaser the qualities of the company product. Large companies have their own publicity department, whereas smaller manufacturing units usually engage advertising agencies and publicity organisations. Creative graphical work plays an important part of the task in advertising and publicity, and close liaison between publicity and technical departments is essential.

Finance

It will not be a case of wandering from the essentials of this chapter if a little consideration is given to the important subject of finance within a manufacturing organisation. The investment in a company can only be recouped when sufficient products are being purchased by the users. Deduct the costs involved for design, development, marketing and manufacture, from the sales achieved, and the result hoped for is the recovery of investment. However, it is the overall profit that is of interest to those who invest money, remembering that investments can be a risky business! The product may sell well initially, and give short-term profits, only for sales to fall away and the product to become a poor investment. It is now perhaps worth while to reflect for a moment upon the complexities of design, development, marketing and manufacture of a new motor vehicle or similar product, remembering that they are all variables. As a summary, four financial stages have been considered (see figure 6.15).

It will be gathered from what has been mentioned that function, symmetry and proportion, sales, service and customer liaison all depend upon *decision-making*. This can be defined as the ability to make decisions in the face of uncertainty, but with a full grasp of all the factors involved. The uncertainty arises because all decisions that are made are with regard to the future. To reach a particular

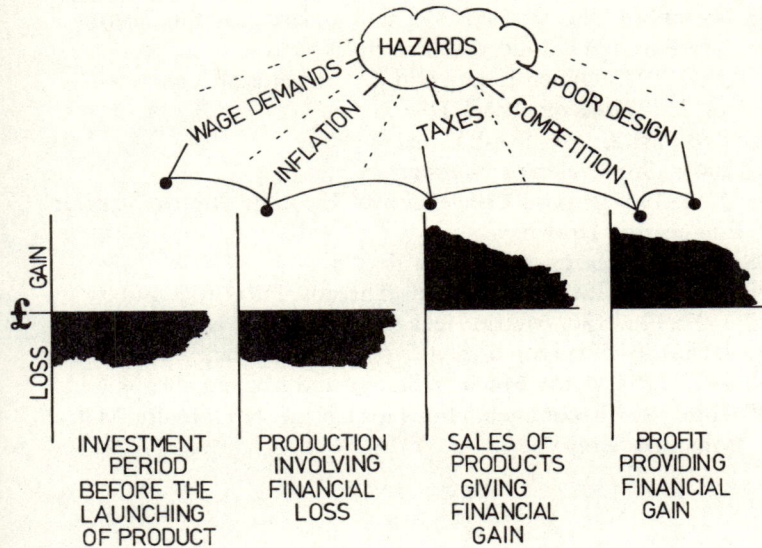

Figure 6.15

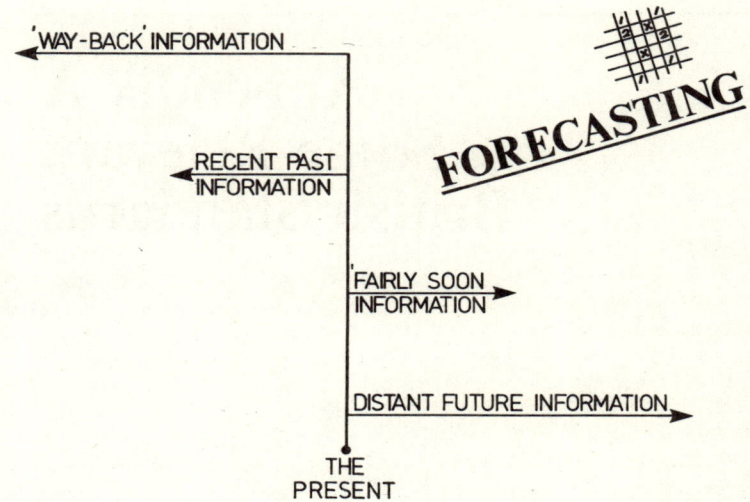

Figure 6.16

decision we must forecast, and so forecasting helps us to make decisions. Both design and marketing require the use of forecasting and, in simple terms, forecasting can be illustrated (see figure 6.16).

Once again, information is the essential ingredient for the success of a manufacturing unit. The history and background of company products in the form of statistical records embracing systems for recording design data, costing and sales, for past and current products, is essential for achieving feasibility of new designs.

There are people who devote all their working lives designing equipment that is rarely considered by the mass of the population. For example, sophisticated mining machinery, complex gearboxes, power-station equipment and precision bearing units that help industrial and domestic products to run smoothly. There is packaging machinery, food preparation machinery, bottling and capping equipment and endless other items that provide small acclaim for the design team involved. What then is good design?

What makes a good designer? We may start to think in terms of functional and aesthetic appreciation and in no time at all begin to fill pages in an attempt to answer the questions posed. But stated simply, and with the prospect of being accused of understating the role of the designer, perhaps the modesty often found within a design group is catered for in the definition at the beginning of this book.

Appendix A
Some Relevant
British Standards

BS Handbook No. 18: 1966 Metric Standards for Engineering
BS 308: Part 1: 1972 General Principles
    Part 2: 1972 Dimensioning and Tolerancing of Size
    Part 3: 1972 Geometrical Tolerancing
BS 499: Part 2: 1965 Symbols for Welding
BS 2045: 1965 Preferred Numbers
BS 2856: 1957 Precise Conversion of Inch and Metric Sizes on
    Engineering Drawings
BS 3429: 1961 Sizes of Drawing Sheets
BS 3643: 1967 ISO Metric Screw Threads, Parts 1, 2, and 3
BS 3692: 1967 ISO Metric Hexagon Bolts, Screws and Nuts
BS 4168: 1967 Hexagon Socket Screws and Wrench Keys, Metric
BS 4183: 1967 Metric Machine Screws and Machine Screw Nuts
BS 4186: 1967 Recommendations for Clearance Holes for Metric
    Bolts and Screws

Appendix B The Forming of Materials; Properties and Uses

Bending—mild steel, wrought iron, annealed carbon steel, copper, lead, zinc, aluminium.

Casting—lead, aluminium, brass, bronze, cast iron, zinc alloys.

Pressing—mild steel, copper, aluminium, brass.

Extruding—aluminium, copper.

Drawing—steel, copper, silver.

Beating—copper, gilding metal, steel, brass, aluminium.

Forging—wrought iron, mild steel (copper, silver and aluminium can be cold forged).

Machining—brass, bronze, aluminium, mild steel, free cutting steel, annealed carbon steel, cast iron.

Cutting with hand tools—all mentioned previously except hardened carbon steel and case hardened mild steel. Materials like glass can be cut with tipped tools or ground to shape with abrasives.

Material	Properties and uses
Aluminium Al Pure	High thermal conductivity and corrosion resistance; utilised for cooking utensils.
	High ductility and malleability; utilised for disposable tube containers and cooking foil.
	High electrical conductivity for electric cables.
Alloyed with magnesium rolled, not heat treated.	Much stronger and used in sheet form for containers and panelling in the transport industry.
Alloyed with magnesium, copper and other trace elements, rolled and heat treated.	As strips and rolled sections, used for structural work in domestic window frames, aircraft industry, etc., e.g. Duralumin.
Alloyed with silicon and cast.	Sand casting, gravity and pressure die-casting, good flow properties in casting.
Antimony Sb	Used as an alloying element in bearings and type metal.

Material	Properties and uses	Material	Properties and uses
Beryllium Be	A very light metal, very expensive to refine, used to harden copper and in nuclear engineering.	Gunmetal: 85–90% Cu, 10% Sn and 2–5% Zn	Bearings and corrosion-resistant castings.
Bismuth Bi Alloyed with lead and tin to form low melting point alloys.	Plugs for automatic fire extinguishers, dental fillings. Melting point can be reduced to below that of boiling water by further alloy additions.	Cupro-nickel: 30–80% Cu, 70–20% Ni and 0.4% Mn	Coins and chemical plant because of corrosion resistance.
		Nickel-silver: 60% Cu, 18% Ni and 22% Zn	Cutlery
Cadmium Cd	As protective plate and alloyed with copper to increase strength of telephone wires.	*Germanium* Ge	Transistors and rectifiers
		Iron Fe Pure	Pure iron is soft and, in its pure state, almost useless.
Chromium Cr	Alloyed in steel to increase hardness and resistance to corrosion (causes brittleness). Used for decorative protective plating.	Mild steel containing up to 0.25% C	Nails, motor car bodies, structural steel and free machining steel.
		Medium carbon steel containing 0.25–0.6% C	Steel tubes, forgings, gears, rails, wire ropes and hand tools.
Cobalt Co	Alloyed in steel to make permanent magnets and high-speed cutting steel for tools.	High carbon steel containing 0.6–1.5% C	Tools and cutting blades.
		Cast iron: 3–4% C, 1–3% Si and 0.5–1% Mn	Castings, including car engine blocks.
Copper Cu Pure	High thermal conductivity and corrosion resistance utilised in radiators, boilers and heating equipment. High electrical conductivity and ductility utilised for electrical conductors and wires.	Nickel steel: 0.2% C, 0.4% Mn and 4.0% Ni	Gears—can be surface or case hardened.
		Manganese steels: 0.35% C and 1.50% Mn	Cheap substitute for nickel chromium steel in automobile engineering.
Brass: 90% Cu, 10% Zn 60–70% Cu, 40–30% Zn	Gilding metal used for jewellery. The higher the copper content the greater its ductility. 60–40 brass with 3% Pb has good machining qualities, with 1% Sn it has good corrosion resistance (naval brass).	Tool steel: 0.35% C, 5.00% Cr, 1.00% Mo, 0.40% V and 1.35% W	Press tools, punches and dies for forging and extrusion.
		High speed steel: 0.8% C, 4.0% Cr, 18.0% W and 1.5% V	Lathe tools, drills, taps, dies
		Stainless steel: 0.3% C, 0.5% Mn and 13% Cr	Tools, springs, cutlery
Phosphor bronze: 85–95% Cu, 15–5% Sn and 0.5% P	Springs, bearings	Decorative stainless steel (18/8): 0.1% C, 0.8% Mn, 8.0% Ni and 18.0% Cr	Domestic and decorative purposes.

Material	Properties and uses
Lead Pb	
90–99% Pb and 10–1% Sb	Accumulators, cable sheathing and pipes, and roof covering.
80% Pb, 13% Sb and 7% Sn	Printing type
Magnesium Mg	
Alloyed with Al, Zn and Th	Very light castings for aircraft industry, not very high strength.
Manganese Mn	Used as a deoxidiser and de-sulphuriser in steel.
Mercury Hg	Scientific uses.
Molybdenum Mo	Reduces brittleness in nickel–chromium steels, improves high-temperature strength.
Nickel Ni Pure	Toughens steel. Used as corrosion-resistant plating and a base for chromium plating.
65–80% Ni, 15–20% Cr and 20% Fe	Electrical heater elements
34% Ni, 4% Cr and 62% Fe	Resistance wire
Nimonic: 60% Ni, 20% Cr, 20% Co and 2% Ti	Gas turbine blades
NiFe: 50% Ni, 50% Fe	Thermostats
Platinum Pt	Because of corrosion resistance, used in scientific apparatus and jewellery.
Silicon Si	Heat-resistant steels
Silver Ag	Electrical contacts and jewellery.
Tin Sn Pure	Protective coating on steel (tin plate)
Babbitt metal: 80% Sn, 10% Sb, 5% Cu and 5% Pb	Heavy-duty bearings

Material	Properties and uses
Titanium Ti	Very light, very strong and very corrosion resistant—used in the aerospace industry.
Tungsten W	Used in electric lamps because of its high melting point. It raises the softening temperature of steels.
Vanadium V	It raises the softening temperature of hardened steels.
Zinc Zn Pure	Galvanising.
Die-casting alloys: 95% Zn, 4% Al, 1% Cu	Excellent die-casting properties because of its low melting point. Reasonably strong.

Appendix C
Preferred Metric Basic Sizes;
Preferred Numbers

METRIC BASIC SIZES

Once the decision is reached and accepted regarding the component sizes, a demand is made upon various departments for materials, tools and gauging equipment. Popular sizes of drills, reamers, gauges and standard fasteners are often held in stock, and the obvious extension to this positive approach is to use preferred sizes when detail drawings are being prepared.

The following information has been extracted from Table 1, BS 4318:1968, Preferred Metric Basic Sizes for Engineering. It is reproduced by permission of the British Standards Institution, 2, Park Street, London W1Y 4AA, from whom copies of the complete standard may be obtained.

Preferred basic sizes above 300 mm and up to 1000 mm and above 1000 mm are also dealt with in this standard.

The use of simple numbers is worthy of note, together with the three-way priority headings (see table opposite). Although design features, tool stocks, and similar factors are key considerations in the triple choice, the desired variety reduction can only be fully achieved when first choice sizes are used.

The many standards for sizes of raw materials, bearings, drills, fasteners, etc., are readily accepted, and the same willingness should be shown when finalising component detail. Metrication means more than unit measurement changes. It provides the opportunity for fresh approaches in many sectors of industry, not the least of which is the preferred size approach to the 'shaft running in the bush that fits into the housing'.

PREFERRED NUMBERS

Metrication provides ideas and systems which, although contrary to accustomed and traditional methods, must be accepted. The coherent system of units of measurement involves the presentation of numerical values and also, until SI units are adopted completely, the conversion from one system to another. During the changeover period, the opportunity arises to consider certain aspects such as standardisation in the form of a reasonable range of products, resulting in economy of production. This in turn can affect such

Choice			Choice			Choice		
1st	2nd	3rd	1st	2nd	3rd	1st	2nd	3rd
1	—	—	10	—	—	100	—	—
—	1.1	—	—	11	—	—	—	102
1.2	—	—	12	—	—	—	105	—
—	—	1.3	—	—	13	—	—	108
—	1.4	—	—	14	—	110	—	—
—	—	1.5	—	—	15	—	—	112
1.6	—	—	16	—	—	—	115	—
—	—	1.7	—	—	17	—	—	118
—	1.8	—	—	18	—	120	—	—
—	—	1.9	—	—	19	—	—	122
2	—	—	20	—	—	—	125	—
—	—	2.1	—	—	21	—	—	128
—	2.2	—	—	22	—	130	—	—
—	—	2.4	—	—	23	—	—	132
2.5	—	—	—	—	24	—	135	—
—	—	2.6	25	—	—	—	—	138
—	2.8	—	—	—	26	140	—	—
3	—	—	—	28	—	—	—	142
—	—	3.2	30	—	—	—	145	—
—	3.5	—	—	32	—	—	—	148
up to 10 mm			up to 100 mm			up to 300 mm		

functions as after-sales maintenance together with resultant costs.

It is essential to consider preferred series of numbers in the interests of efficient and economical manufacture, even though they are not directly associated with SI units. Let us consider the theory presented in 1877 by a French engineer, Colonel Charles Renaud. He expressed the view that the range of values to meet most needs was the range which followed a geometrical progression. Further, the largest size of a series should be 10 times the smallest size. It follows that should the first size of a series be one, then the last will be 10, and if five steps in geometrical progression are required, the common ratio is the fifth root of 10. This is termed the R5 Series. If 10 steps are required, then the common ratio is the tenth root of 10,

and is termed the R10 Series. The Renaud Series can be tabulated as follows:

Series	Common ratio	Successive term increase, %
R5	$\sqrt[5]{10}$	1.58	58
R10	$\sqrt[10]{10}$	1.26	26
R20	$\sqrt[20]{10}$	1.12	12
R40	$\sqrt[40]{10}$	1.06	6
R80	$\sqrt[80]{10}$	1.03	3

Use of the common ratio can obviously give impracticable intermediate values, and rounding-off is necessary. The permissible limits for rounded values being $+1.26\%, -1.01\%$.

Terms thus rounded are doubled every three terms in the R10 Series, every six terms in the R20 Series, and every twelve terms in the R40 Series. Extensions and modifications to these series can easily be achieved. Multiplying or dividing by 10 can extend the series, and additional series can be created by using every second, third or fourth term of a series. The additional series are then termed R5/3, R10/4, etc.

To summarise, the size ranges of products can easily be achieved, and it must be remembered that Preferred Numbers are widely accepted internationally.

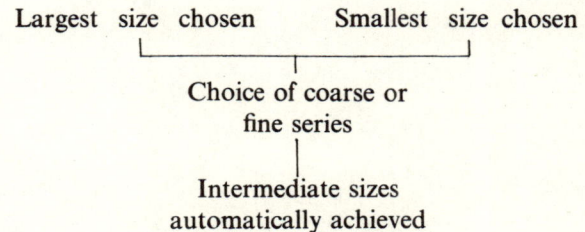

Largest size chosen   Smallest size chosen

Choice of coarse or
fine series

Intermediate sizes
automatically achieved

A detailed treatment of Preferred Numbers is given in BS 2045:1965.

Appendix D
Conversion Tables

Inches to Millimetres

Fractions

in.		mm	in.		mm
$\frac{1}{64}$	0.015 625	0.396 9	$\frac{33}{64}$	0.515 625	13.096 9
$\frac{1}{32}$	0.031 250	0.793 7	$\frac{17}{32}$	0.531 250	13.493 7
$\frac{3}{64}$	0.046 875	1.190 6	$\frac{35}{64}$	0.546 875	13.890 6
$\frac{1}{16}$	0.062 500	1.587 5	$\frac{9}{16}$	0.562 500	14.287 5
$\frac{5}{64}$	0.078 125	1.984 4	$\frac{37}{64}$	0.578 125	14.684 4
$\frac{3}{32}$	0.093 750	2.381 2	$\frac{19}{32}$	0.593 750	15.081 2
$\frac{7}{64}$	0.109 375	2.778 1	$\frac{39}{64}$	0.609 375	15.478 1
$\frac{1}{8}$	0.125 000	3.175 0	$\frac{5}{8}$	0.625 000	15.875 0
$\frac{9}{64}$	0.140 625	3.571 9	$\frac{41}{64}$	0.640 625	16.271 9
$\frac{5}{32}$	0.156 250	3.968 7	$\frac{21}{32}$	0.656 250	16.668 7
$\frac{11}{64}$	0.171 875	4.365 6	$\frac{43}{64}$	0.671 875	17.065 6
$\frac{3}{16}$	0.187 500	4.762 5	$\frac{11}{16}$	0.687 500	17.462 5
$\frac{13}{64}$	0.203 125	5.159 4	$\frac{45}{64}$	0.703 125	17.859 4
$\frac{7}{32}$	0.218 750	5.556 2	$\frac{23}{32}$	0.718 750	18.256 2
$\frac{15}{64}$	0.234 375	5.953 1	$\frac{47}{64}$	0.734 375	18.653 1
$\frac{1}{4}$	0.250 000	6.350 0	$\frac{3}{4}$	0.750 000	19.050 0
$\frac{17}{64}$	0.265 625	6.746 9	$\frac{49}{64}$	0.765 625	19.446 9
$\frac{9}{32}$	0.281 250	7.143 7	$\frac{25}{32}$	0.781 250	19.843 7
$\frac{19}{64}$	0.296 875	7.540 6	$\frac{51}{64}$	0.796 875	20.240 6
$\frac{5}{16}$	0.312 500	7.937 5	$\frac{13}{16}$	0.812 500	20.637 5
$\frac{21}{64}$	0.328 125	8.334 4	$\frac{53}{64}$	0.828 125	21.034 4
$\frac{11}{32}$	0.343 750	8.731 2	$\frac{27}{32}$	0.843 750	21.431 2
$\frac{23}{64}$	0.359 375	9.128 1	$\frac{55}{64}$	0.859 375	21.828 1
$\frac{3}{8}$	0.375 000	9.525 0	$\frac{7}{8}$	0.875 000	22.225 0
$\frac{25}{64}$	0.390 625	9.921 9	$\frac{57}{64}$	0.890 625	22.621 9
$\frac{13}{32}$	0.406 250	10.318 7	$\frac{29}{32}$	0.906 250	23.018 7
$\frac{27}{64}$	0.421 875	10.715 6	$\frac{59}{64}$	0.921 875	23.415 6
$\frac{7}{16}$	0.437 500	11.112 5	$\frac{15}{16}$	0.937 500	23.812 5
$\frac{29}{64}$	0.453 125	11.509 4	$\frac{61}{64}$	0.953 125	24.209 4
$\frac{15}{32}$	0.468 750	11.906 2	$\frac{31}{32}$	0.968 750	24.606 2
$\frac{31}{64}$	0.484 375	12.303 1	$\frac{63}{64}$	0.984 375	25.003 1
$\frac{1}{2}$	0.500 000	12.700 0			

$\frac{1}{1000}$ in.		$\frac{1}{100}$ in.		$\frac{1}{10}$ in.	
in.	mm	in.	mm	in.	mm
0.001	0.025 4	0.01	0.254	0.1	2.54
0.002	0.050 8	0.02	0.508	0.2	5.08
0.003	0.076 2	0.03	0.762	0.3	7.62
0.004	0.101 6	0.04	1.016	0.4	10.16
0.005	0.127 0	0.05	1.270	0.5	12.70
0.006	0.152 4	0.06	1.524	0.6	15.24
0.007	0.177 8	0.07	1.778	0.7	17.78
0.008	0.203 2	0.08	2.032	0.8	20.32
0.009	0.228 6	0.09	2.286	0.9	22.86

Units

in.	mm	10	20	30
0		254.0	508.0	762.0
1	25.4	279.4	533.4	787.4
2	50.8	304.8	558.8	812.8
3	76.2	330.2	584.2	838.2
4	101.6	355.6	609.6	863.6
5	127.0	381.0	635.0	889.0
6	152.4	406.4	660.4	914.4
7	177.8	431.8	685.8	939.8
8	203.2	457.2	711.2	965.2
9	228.6	482.6	736.6	990.6

Millimetres to Inches

Units

mm		10	20	30	40	50	60	70	80	90
0		0.393 70	0.787 40	1.181 10	1.574 80	1.968 51	2.362 21	2.755 91	3.149 61	3.543 31
1	0.039 37	0.433 07	0.826 77	1.220 47	1.614 17	2.007 88	2.401 58	2.795 28	3.188 98	3.582 68
2	0.078 74	0.472 44	0.866 14	1.259 84	1.653 54	2.047 25	2.440 95	2.834 65	3.228 35	3.622 05
3	0.118 11	0.511 81	0.905 51	1.299 21	1.692 91	2.086 62	2.480 32	2.874 02	3.267 72	3.661 42
4	0.157 48	0.551 18	0.944 88	1.338 58	1.732 28	2.125 99	2.519 69	2.913 39	3.307 09	3.700 79
5	0.196 85	0.590 55	0.984 25	1.377 95	1.771 65	2.165 36	2.559 06	2.952 76	3.346 46	3.740 16
6	0.236 22	0.629 92	1.023 62	1.417 32	1.811 03	2.204 73	2.598 43	2.992 13	3.385 83	3.779 53
7	0.275 59	0.669 29	1.062 99	1.456 69	1.850 40	2.244 10	2.637 80	3.031 50	3.425 20	3.818 90
8	0.314 96	0.708 66	1.102 36	1.496 06	1.889 77	2.283 47	2.677 17	3.070 87	3.464 57	3.858 27
9	0.354 33	0.748 03	1.141 73	1.535 43	1.929 14	2.322 84	2.716 54	3.110 24	3.503 94	3.897 64

mm		100	200	300	400	500	600	700	800	900
0		3.937 01	7.874 02	11.811 0	15.748 0	19.685 1	23.622 1	27.559 1	31.496 1	35.433 1
10	0.393 70	4.330 71	8.267 72	12.204 7	16.141 7	20.078 8	24.015 8	27.952 8	31.889 8	35.826 8
20	0.787 40	4.724 41	8.661 42	12.598 4	16.535 4	20.472 5	24.409 5	28.346 5	32.283 5	36.220 5
30	1.181 10	5.118 11	9.055 13	12.992 1	16.929 1	20.866 2	24.803 2	28.740 2	32.677 2	36.614 2
40	1.574 80	5.511 81	9.448 83	13.385 8	17.322 8	21.259 9	25.196 9	29.133 9	33.070 9	37.007 9
50	1.968 51	5.905 52	9.842 52	13.779 5	17.716 5	21.653 6	25.590 6	29.527 6	33.464 6	37.401 6
60	2.362 21	6.299 22	10.236 20	14.173 2	18.110 3	22.047 3	25.984 3	29.921 3	33.858 3	37.795 3
70	2.755 91	6.692 92	10.629 90	14.566 9	18.504 0	22.441 0	26.378 0	30.315 0	34.252 0	38.189 0
80	3.149 61	7.086 62	11.023 60	14.960 6	18.897 7	22.834 7	26.771 7	30.708 7	34.645 7	38.582 7
90	3.543 31	7.480 32	11.417 30	15.354 3	19.291 4	23.228 4	27.165 4	31.102 4	35.039 4	38.976 4

Fractions

$$\frac{1}{1000} \text{ mm}$$

mm	in.
0.001	0.000 039
0.002	0.000 079
0.003	0.000 118
0.004	0.000 157
0.005	0.000 197
0.006	0.000 236
0.007	0.000 276
0.008	0.000 315
0.009	0.000 354

$$\frac{1}{100} \text{ mm} \qquad \frac{1}{10} \text{ mm}$$

mm	in.	mm	in.
0.01	0.000 39	0.1	0.003 94
0.02	0.000 79	0.2	0.007 97
0.03	0.001 18	0.3	0.011 81
0.04	0.001 57	0.4	0.015 75
0.05	0.001 97	0.5	0.019 67
0.06	0.002 36	0.6	0.023 67
0.07	0.002 76	0.7	0.027 56
0.08	0.003 15	0.8	0.031 50
0.09	0.003 54	0.9	0.035 43

Assignments

(1) Details are shown of part of a rocker mechanism (see figure A1). Provide

    (a) an assembled sectional front elevation
    (b) an assembled sectional end elevation
    (c) a separate parts list
    (d) a set of detail drawings.

Work to the following instructions:

    (i) A headed phosphor bronze bush to be fitted at each end of the body.

    (ii) Overall dimensions to be maintained, but modify any other features to suit.

    (iii) The body detail is to be cast, the two arms are to be fabricated.

    (iv) The 25 dimension on the arm detail is the centre of a fixing pin to fasten the arm to the shaft. Show the fixing method.

    (v) Faces marked B on the arm and body, and the base of the body, are to be machined.

    (vi) Show a 38 diameter shaft protruding 80 at each end through the arms with a conventional break. All dimensions in millimetres.

(2) An outline section of a valve is shown (see figure A2). The body is cast and the valve spindle passes through a bridge which is carried by the top cover. The spindle then passes through a gland/packing and then connects with the replaceable valve seat which is located in the body. The body is flanged, and is connected by bolts into a pipeline which carries water, steam, etc. For smaller sizes, the body can be internally or externally threaded for connection into the line.

The arrangement offers no aesthetic appeal, and yet functionally a similar valve is fitted into domestic heating systems which are on view in carpeted and furnished rooms. The designer must achieve, therefore, both a functionally correct design (easy on–off operation, no leakages, etc.), and provide aesthetic qualities which must verge on the unobtrusive. After all, a radiator valve in a domestic lounge should not give the impression of an oil rig pipeline!

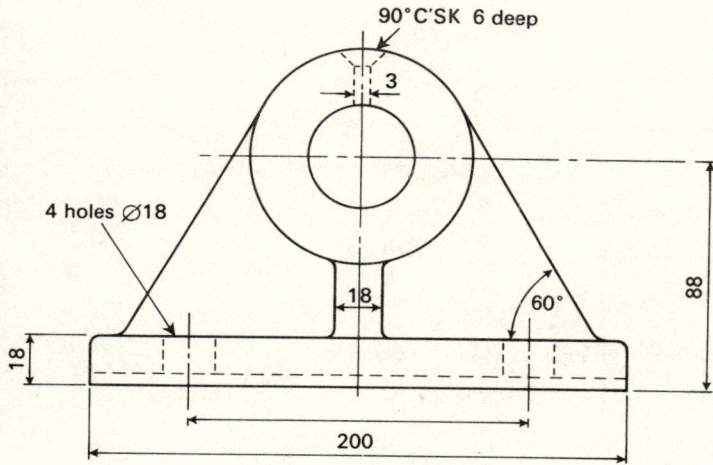

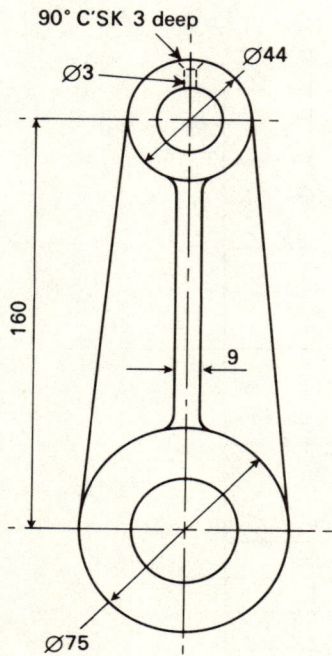

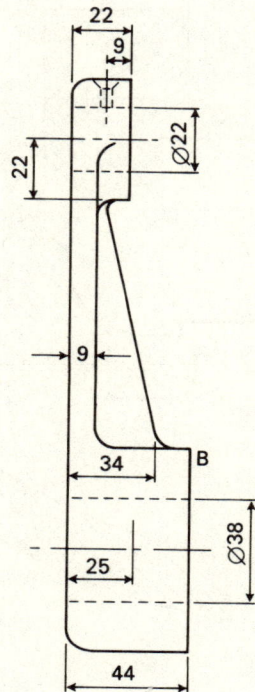

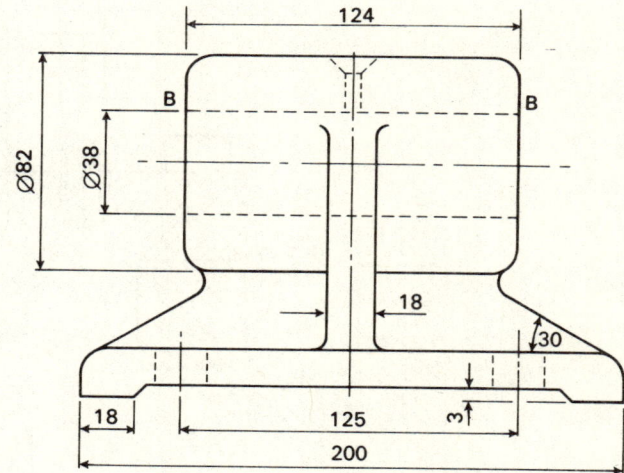

90°C'SK 6 deep

3

4 holes ⌀18

18

60°

18

88

200

90° C'SK 3 deep

⌀3

⌀44

⌀22

22

22

9

160

9

9

34

B

25

⌀38

⌀75

44

124

B

B

⌀82

⌀38

18

30

18

125

3

200

Figure A1

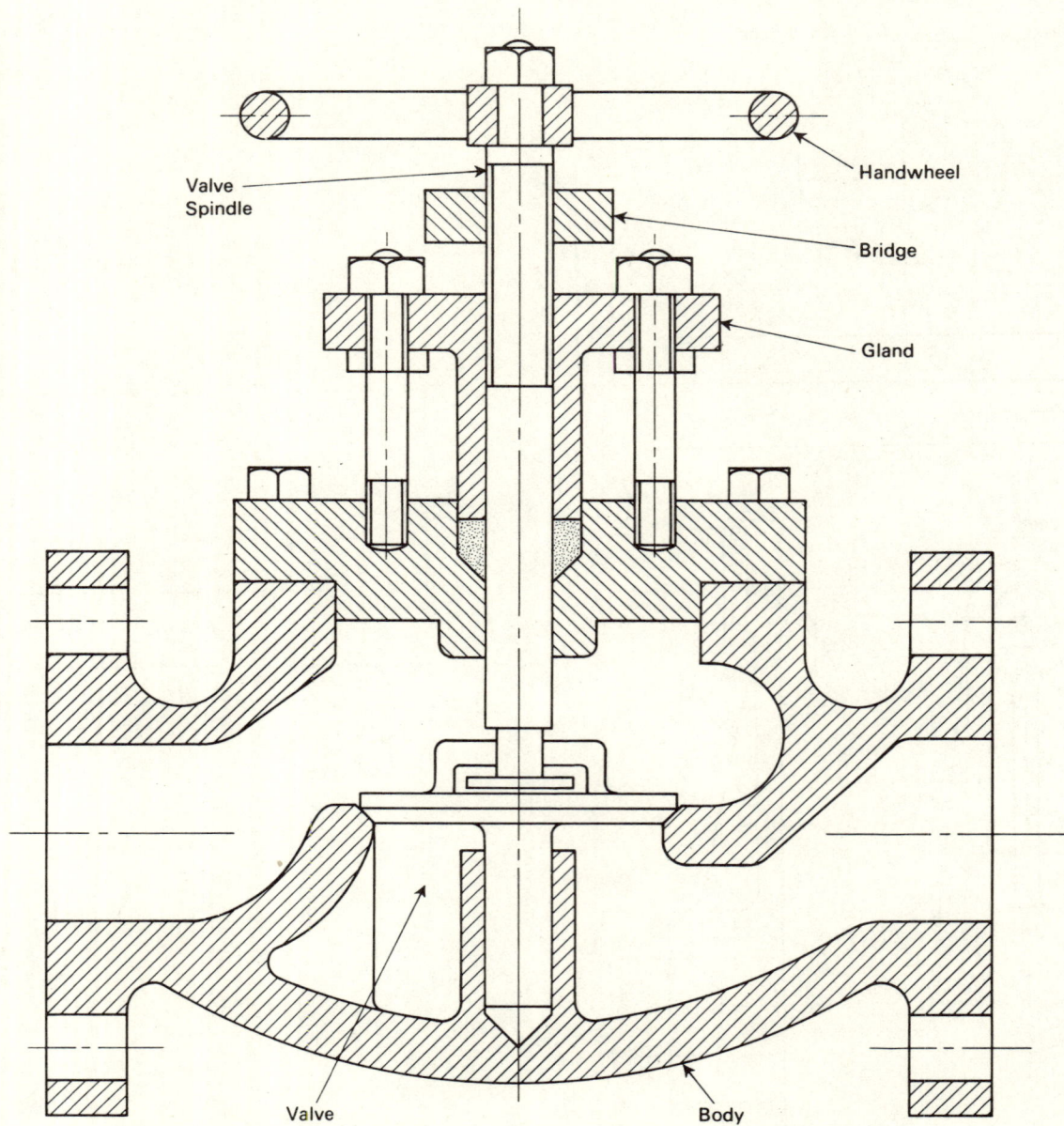

Handwheel

Valve
Spindle

Bridge

Gland

Valve

Body

Figure A2

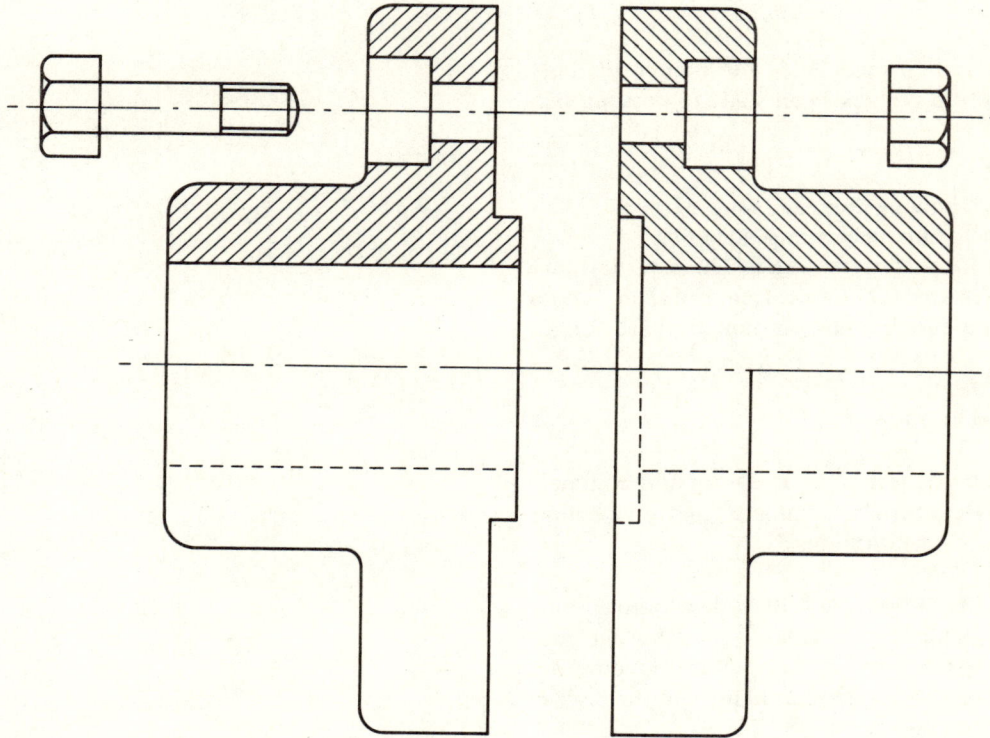

Figure A3

Make neat design sketches of the functional aspects of such a small domestic valve, and also various alternative external designs which will provide aesthetic appeal. What materials would you use? What difficulties can you find in combining aesthetic and functional qualities? Present your sketches and written findings in a neat manner suitable for discussion at management level.

(3) The two halves of a shaft coupling are shown in basic outline (see figure A3). This is a rigid coupling, which can be defined as a device for connecting two shafts in such a manner that no displacement of one relative to the other can occur, the two shafts behaving as one.

Information

The two halves are located by a spigot diameter.

There is a parallel keyway cut in each half.

There are six equally spaced turned steel bolts positioned in reamed holes.

Bore diameter 25 mm (use H7 hole from tables in BS 4500).

Depth of keyway from periphery of hole on centreline 28.5/28.3 mm.

Width of keyway 6.38/6.35 mm.

Bolt diameter 8 mm.

Overall length assembled 120 mm.

Flange diameter not to exceed 120 mm.

Design Considerations

Turned circular recess and reamed holes for bolt heads/nuts. Provision for both halves to be locked on shaft to prevent axial movement.

Provide the following

(a) An arrangement drawing complete with parts list.
(b) Detail drawings for manufacturing purposes.
(c) Since couplings are part of rotating machinery, they must be adequately guarded. Assume that the centreline of shaft to floor level is 180 mm; design a suitable floor mounted guard for the coupling.
(d) A product specification.
(e) A functional specification.

(4) Consider a domestic cooker of your choice and comment upon the features of order, variety, symmetry and proportion. Sketches should be used in your appraisal.

(5) Obtain a new car brochure from a local distributor. Study the information provided and make sketches to show how, in your opinion, extra stowage space, improved dashboard layout, etc., could be achieved. Your modifications should not put the price of the car into a higher price bracket.

(6) A towel rail is to consist of a swinging cantilever arm, which will be wall fixed. Construct a full-sized model and prepare the relevant sketches and information which will enable you to finalise the design. Discuss the main design consideration in detail.

(7) 'The more produced, the lower the cost.' Discuss the merits of this statement using everyday products in your argument.

Questions

(1) Sketch freehand a section taken on the centreline of a double thrust ball-bearing. Your sketch should be clearly labelled. State where such a bearing is used.

(2) Four-way tool posts are used on centre lathes. Prepare a freehand sketch of the indexing mechanism.

(3) In good proportion, sketch an arrangement showing how the adjustment of the vee-slide of a compound slide on a lathe is achieved.

(4) Explain, using sketches where necessary, why tapered bearings are used on certain engineering assemblies. Draw a section, in good proportion, of a rotating tailstock centre.

(5) Using sketches, explain what is meant by 'accumulation of tolerances'.

(6) Using the latest BS 308: 1972 recommendations, show

  (a) the application of auxiliary dimensions
  (b) holes equally spaced on a pitch circle
  (c) the use of ordinates for holes lying on a pitch circle.

(7) Describe, with the aid of sketches the BS 308 method of dimensioning

  (a) a keyway in a shaft
  (b) a counterbored hole
  (c) a spot faced hole
  (d) a machined surface
  (e) holes unequally spaced on a circular pitch.

(8) Sketch the BS 308 convention for

  (a) a roller bearing
  (b) a splined shaft
  (c) a spur gear.

(9) Sketch and describe two patented types of locknut.

(10) Many houses are warmed by central heating. Using sketches, describe the construction and operation of a thermostat for controlling the room temperature.

(11) Explain, using sketches, the difference between double and treble riveting.

(12) Describe, using sketches, how a riveted joint can fail. Give reasons for such failures and suggest how they can be avoided.

(13) Two steel plates, 125 mm wide and 15 mm thick, are to be riveted together using two cover straps. Produce a drawing of the arrangement showing the spacing of the rivets. State the type of rivets you have used.

(14) Assuming a house is centrally heated by a solid fuel boiler and that the cold water is supplied from a roof tank, sketch the arrangement that would be suitable for a piping arrangement for a house with two rooms and a bathroom upstairs and three rooms on the ground floor. Label your diagram.

(15) Two tie bars, 200 mm wide and 12 mm thick, are to be joined using a double butt strap riveted joint and 22 snap headed rivets. Produce a suitable sketch of the arrangement.

(16) Draw in accordance with BS 308 the conventional representation of

  (a) a compression spring
  (b) a tension spring
  (c) an internal thread
  (d) an external thread.